SpringerBriefs in Applied Sciences and Technology

SpringerBriefs present concise summaries of cutting-edge research and practical applications across a wide spectrum of fields. Featuring compact volumes of 50–125 pages, the series covers a range of content from professional to academic.

Typical publications can be:

- A timely report of state-of-the art methods
- An introduction to or a manual for the application of mathematical or computer techniques
- A bridge between new research results, as published in journal articles
- A snapshot of a hot or emerging topic
- An in-depth case study
- A presentation of core concepts that students must understand in order to make independent contributions

SpringerBriefs are characterized by fast, global electronic dissemination, standard publishing contracts, standardized manuscript preparation and formatting guidelines, and expedited production schedules.

On the one hand, **SpringerBriefs in Applied Sciences and Technology** are devoted to the publication of fundamentals and applications within the different classical engineering disciplines as well as in interdisciplinary fields that recently emerged between these areas. On the other hand, as the boundary separating fundamental research and applied technology is more and more dissolving, this series is particularly open to trans-disciplinary topics between fundamental science and engineering.

Indexed by EI-Compendex and Springerlink.

More information about this series at http://www.springer.com/series/8884

Reggie Davidrajuh

Modeling Discrete-Event Systems with GPenSIM

An Introduction

 Springer

Reggie Davidrajuh
Department of Electrical Engineering
  and Computer Science
University of Stavanger
Stavanger
Norway

Additional material to this book can be downloaded from http://extras.springer.com.

ISSN 2191-530X   ISSN 2191-5318   (electronic)
SpringerBriefs in Applied Sciences and Technology
ISBN 978-3-319-73101-8   ISBN 978-3-319-73102-5   (eBook)
https://doi.org/10.1007/978-3-319-73102-5

Library of Congress Control Number: 2017964442

MATLAB® is a registered trademark of The MathWorks, Inc., 1 Apple Hill Drive, Natick, MA 01760-2098, USA, http://www.mathworks.com.

Printed on acid-free paper

This Springer imprint is published by Springer Nature
The registered company is Springer International Publishing AG
The registered company address is: Gewerbestrasse 11, 6330 Cham, Switzerland

This book is dedicated to my brothers and sisters:

Daisy Vijayaraghavan (Singapore)

Rhavy Davidrajuh (London, UK)

Antonyroy Davidrajuh (Manipay, Sri Lanka)

Lucky Valentine (Colombo, Sri Lanka)

Preface

Petri net is being widely accepted by the research community as a tool for modeling and simulation of discrete-event systems. There are a number of Petri net tools available for academic and commercial use. These tools are advanced tools powerful enough to model complex and large systems. In this book, a new Petri net simulator called General Purpose Petri net Simulator (GPenSIM) is introduced. GPenSIM runs on MATLAB platform. GPenSIM is designed with three specific goals: (1) modeling, simulation, performance analysis, and control of discrete-event systems, (2) a tool that is easy to use and extend, and (3) allowing Petri net models to be integrated with other MATLAB toolboxes (e.g., Fuzzy Logic, Control Systems).

This book is written as a book to assist students and practicing engineers develop mathematical models of discrete-event systems with GPenSIM. Thus, Petri net theory is kept to a minimum, and it is expected that the students know some Petri net theory beforehand. To help the readers grasp the modeling basics quickly and easily, there are many examples worked out in this book. These examples are simple and easy to follow. Both the simulator GPenSIM and codes for examples (M-files) can be downloaded from the companion website: http://www.davidrajuh. net/gpensim.

This book is based on the new version of GPenSIM, version 10. The version 10 is the first stable version of GPenSIM, which is thoroughly checked for bugs. In addition, there are some changes in input and output parameters of some of the GPenSIM functions. Due to these changes, unfortunately, version 10 is not upward compatible with its previous versions.

This book was written while I was staying at the Silesian University of Technology, Poland, for my sabbatical leave. I want to thank the Institute of Engineering Processes Automation and Integrated Manufacturing Systems of the Faculty of Mechanical Engineering, for hosting me, especially Professor Dr. hab. inz Bozena Skolud and

Dr. hab. inz Damian Krenczyk for all their help during my sabbatical stay in Poland. Finally, I want to thank my wife Ruglin and my daughter Ada for their love and patience with me.

I hope you enjoy the book!

Stavanger, Norway   Reggie Davidrajuh
October 2017

Contents

Chapter 1
Introduction

This chapter introduces Petri nets and the General Purpose Petri net Simulator (GPenSIM). Petri net is widely accepted for modeling and simulation of discrete-event systems, due to its graphical representation and the well-defined semantics. GPenSIM defines a Petri net language on MATLAB platform. GPenSIM is also a simulator with which Petri net models can be developed, simulated, and analyzed. GPenSIM is easy to learn, use, and extend.

1.1 What Is GPenSIM?

GPenSIM defines a Petri net language for modeling and simulation of discrete-event systems on MATLAB platform. GPenSIM is developed by the author of this book. GPenSIM is also a simulator with which Petri net models can be developed and simulated. In addition, GPenSIM can also be used as a real-time controller. Even though GPenSIM is a new simulator, it is being used by many researchers around the world. Recently, a team of Australian researchers chose GPenSIM as the ideal tool for modeling and simulation of marked graph (a Petri net class), because of GPenSIM's flexibility in being able to control the system via the model (Cameron et al. 2015). For the same reason, Tilbury laboratory at the University of Michigan also selected GPenSIM as the tool for modeling manu-facturing system (Lopez et al. 2017).

GPenSIM supports many Petri nets extensions, such as inhibitor arcs, transition priorities, enabling functions, color extension. In addition, it provides a collection of functions for performance analysis. Because of its flexibility, it is also easy to implement any other Petri net extensions with GPenSIM, e.g., Attributed Hybrid Dynamical net (Lopez et al. 2017), Cohesive Place-Transition Nets with inhibitor arcs (Davidrajuh and Saadallah 2016).

© The Author(s) 2018   1
R. Davidrajuh, *Modeling Discrete-Event Systems with GPenSIM*, SpringerBriefs in
Applied Sciences and Technology, https://doi.org/10.1007/978-3-319-73102-5_1

1.2   Petri nets: Basic Concepts

This section introduces the basic concepts of Petri nets and its relation to time. This section gives only a brief introduction. For a formal study of Petri nets, interested readers are referred to books like Cassandras and Lafortune (2009), DiCesare et al. (1993), Murrata (1989), and Peterson (1981).

1.2.1   P/T Petri net

A P/T Petri net contains two types of elements: places and transitions; places generally represent passive elements (such as input and output buffers, conveyor belts) and transitions represent active elements (such as machines, robots). Petri net is a directed bipartite graph meaning a place can only be connected to transition(s) and a transition to place(s); the connections between places and transitions are termed as arcs.

   In addition to places, transitions, and arcs, Petri net also has tokens. Tokens represent objects that can flow around in a network of nodes, e.g., materials in a material flow system, data (or information) in an information flow. Places hold tokens; tokens move from place to place via the arcs. Tokens are shown as black spots in a Petri net. If a place has a large number of tokens, then it is customary to show the number of tokens with numerals than black spots.

   The arcs that connect places to transitions and transitions to places have the default weight of one. If an arc has a weight that is greater than unity, then the weight is shown in the arc. The arc weight represents the capacity of the arc to transport a number of tokens simultaneously at a time.

   Figure 1.1 shows three places p_1, p_2, and p_3. These three places hold 4, 3, and 1 tokens, respectively. When a transition fires, a number of tokens are taken ('consumed') from the input places and new tokens are deposited ('produced') into the output places; the arc weights determine the number of tokens consumed and produced. For a transition to be able to fire, the number of tokens in the input places must be equal or larger than the weights of the arcs connecting the input places to the transition. The transition will then become able to fire (*enabled transition*). Figure 1.2 shows the state of the sample Petri net from Fig. 1.1 after the transition t_1 has fired once.

Fig. 1.1 Sample Petri net

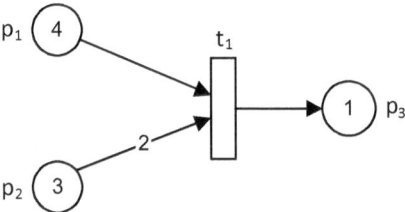

Fig. 1.2 Petri net after one
firing of t_1

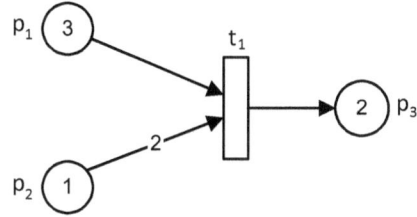

1.2.2 Formal Definition of P/T Petri nets

A P/T Petri net is a 4-tuple (P, T, A, m_0)
where
 P is the set of places, $P = \{p_1, p_2, \ldots, p_n\}$,
 T is the set of transitions, $T = \{t_1, t_2, \ldots, t_m\}$,
 A is the set of arcs (from places to transitions and from transitions to places),

$$A \subseteq (P \times T) \cup (T \times P), \text{ and}$$

m is the row vector of markings (tokens) on the set of places

$$m = [m(p_1), m(p_1), \ldots, m(p_n)] \in \mathbf{N}^n, \ m_0 \text{ is the initial marking.}$$

1.2.3 Input and Output Places of a Transition

In the Petri net shown in Fig. 1.2, the places p_1 and p_2 are input places to transition
t_1, and p_3 is an output place of transition t_1. It is convenient to use $I(t_j)$ to represent
the set of input places to transition t_j and $O(t_j)$ to represent the set of output places to
transition t_j when describing a Petri net:

$$I(t_j) = \{p_i \in P : (p_i, t_j) \in A\}$$
$$O(t_j) = \{p_i \in P : (t_j, p_i) \in A\}$$

We see from Fig. 1.1 that the weight of the arc from input place p_2 to transition
t_1 has a weight of two. This is denoted by $w(p_2, t_1) = 2$.

1.2.4 Enabled Transitions

A transition $t_j \in T$ in a Petri net is said to be *enabled* if (Cassandras and Lafortune
2009): $m(p_i) \geq w(p_i, t_j)$ for all $p_i \in I(t_j)$.

The transition t_1 in Fig. 1.1 is enabled, since the numbers of tokens in the input places p_1 (4) and p_2 (3) are at least as large as the weight of the arcs connecting them to t_1 ($w(p_1, t_1) = 1$ and $w(p_2, t_1) = 2$).

1.3   Petri net Dynamics

The markings of a Petri net, which is the distribution of tokens to the places, represent the state of the Petri net. A Petri net representing a discrete-event system, where the transitions represent events, goes through many states during a simulation process. The different states could be represented by the row vector of markings (the 4th tuple):

$$m = [m(p_1), m(p_2), \ldots, m(p_{n1})].$$

The number of states an *infinite capacity net* can have is generally infinite since each place can hold an arbitrary non-negative integer number of tokens (Murata 1989). A *finite capacity net*, on the other hand, will have a given number of possible states.

The *state transition function*, $f : \aleph^n \times T \rightarrow \aleph^n$, of a Petri net is defined for a transition $t_j \in T$ if and only if, $m(p_i) \geq w(p_i, t_j)$ for all $p_i \in I(t_j)$.

If $f(m, t_j)$ is defined, then $m' = f(m, t_j)$, where

$$m'(p_i) = m(p_i) - w(p_i, t_j) + w(t_j, p_i), \quad i = 1, \ldots, n.$$

1.3.1   *Time and Petri net*

P/T Petri net is untimed. In P/T Petri net, all the transitions possess zero firing time, like Dirac delta function (impulse). Since transitions in P/T Petri net take zero time to fire, they are called 'primitive transitions' (Peterson 1977), as they cannot represent any real event which always takes time. Though GPenSIM can be used for P/T Petri nets, it is mainly for Timed Petri nets. In Timed Petri nets, *all the transitions* are considered to be 'non-primitive,' thus have finite (nonzero) firing times; maybe some of the transitions have very small firing times but are not zero.

In GPenSIM, if firing times are not assigned to any of the transitions, then the system is assumed as a P/T Petri net. It is not acceptable to assign firing times to some of the transitions and let the other transitions take zero value; in other words, a system can be either P/T Petri net (all transitions are untimed) or Timed Petri net where all the transitions are assigned finite firing times (not zero).

1.3.2 Coverability Tree

There are tools to analyze behavioral properties of a system, including:

- Reachability, Boundedness, Conservativeness, Liveness, and Reversibility.

The coverability tree is one of the tools that can be used to analyze the properties of a Petri net; coverability tree consists of a tree of markings and possible transitions between. The coverability tree can be finite (with or without 'omega') or infinite. An infinite coverability tree is unbounded.

1.4 A Simple P/T Petri net Model

The simple Petri net shown in Fig. 1.3 is a model for business logic computation. The computation takes two database records and one business rule and produces one business decision. In a Petri net, sources (like business rules and database records) and outputs (like business decisions) are called places, drawn as circles (e.g., p_1). Computations (or events) are called transitions, drawn as thin rectangular boxes (e.g., t_1). An arc connects a place to a transition, or a transition to a place, representing a path for a discrete part to flow. A place usually holds a number of parts, like database records. The number of parts inside a place is indicated by the tokens (e.g., n_1 in p_1 and n_2 in p_2), either as an integer or by a number of black spots equal to the integer).

1.5 Timed Petri net

GPenSIM interprets Timed Petri net as follows:

- **No variable duration of events**: The transitions representing events are assigned firing time beforehand. The pre-assigned firing time can be deterministic (e.g., firing time $dt = 5$ TU) or stochastic (e.g., firing time dt is normally distributed with mean value 10 TU and standard deviation 2 TU ('normrnd (10,2)'). However, variable firing times are not possible.

Fig. 1.3 Petri net model for business logic computations

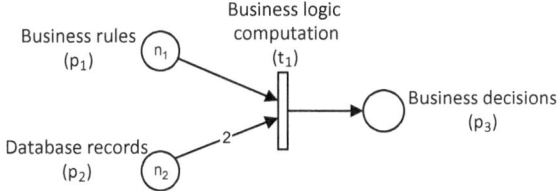

- Maximal-step firing policy: Timed Petri net operates with the maximal-step firing policy, meaning if more than one transition is collectively enabled and are not in conflict with each other at a point of time, then all of them fire at the same time (Popova-Zeugmann 2013).
- *Enabled* transition starts *firing* immediately: This means, there is no (forcibly) induced delay between the time a transition becomes enabled and the time it is allowed to fire.

1.6 Atomicity and Virtual Tokens

Figure 1.4 explains how non-primitive transition t_i of a Timed Petri net (firing time of the transition is not zero) can be understood in terms of primitive (firing time is zero) transitions of P/T Petri nets. As Fig. 1.4 shows, each non-primitive transition in Timed Petri net can be considered as an assembly of four elements. It has two primitive transitions *starter* and *stopper*, and a *virtual place* between them. In addition, there is a place *pme* with an initial token (pme: the place to impose mutual exclusion) in order to make sure that once the starter has fired, it will not fire again until the stopper is fired.

Figure 1.5 explains the firing of t_i. Whenever t_i is ready to fire, *starter* fires immediately and passes the input tokens into the virtual place; the input tokens will stay in the virtual place for an amount of time (delay) equal to the firing time of t_i. At the completion of the delay, the stopper fires immediately, consuming all the

Fig. 1.4 Composition of a non-primitive transition

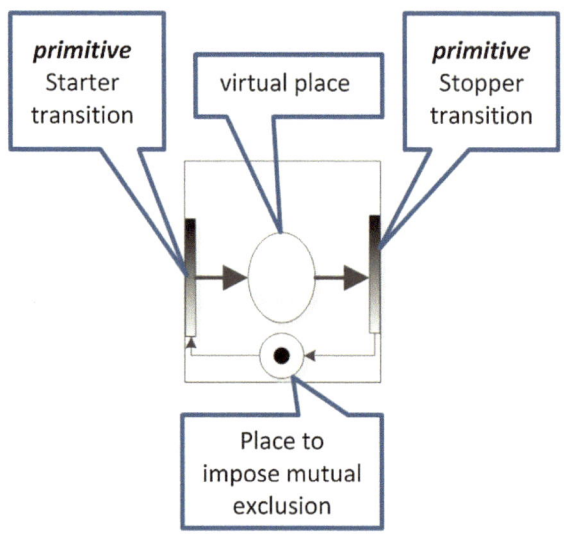

virtual tokens and depositing newer output tokens into the output places. The firing mechanism described above makes sure that the tokens are accountable (do not disappear) anytime during the firing of the non-primitive transition t_i. Thus, *atomicity* property is upheld.

Fig. 1.5 Maintaining the 'atomicity' property during the firing of a non-primitive transition

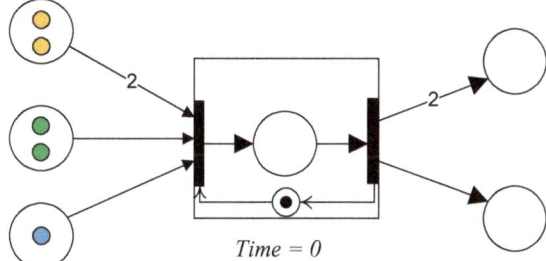

2a) Non-primitive transition is enabled and about to fire

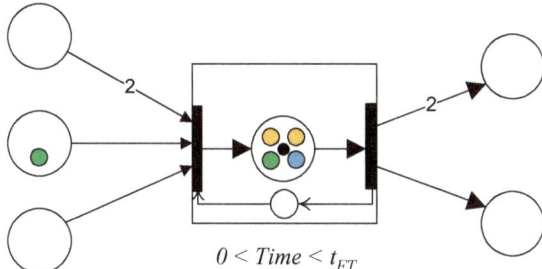

2b) Non-primitive transition starts firing: primitive starter fires instantly, removing the tokens from the input place and placing them as virtual tokens inside the virtual place

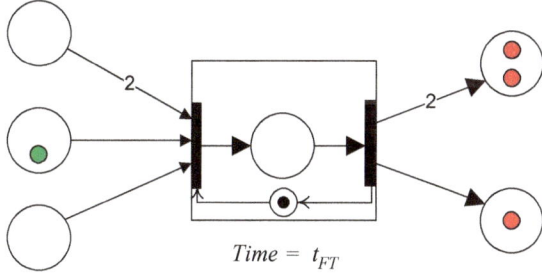

2c) Non-primitive transition completes firing: primitive stopper transition fires instantly, removing the virtual tokens from the virtual place and depositing output tokens into the output place

1.7   Defining Timed Petri net (TPN)

Timed Petri nets (TPN) are ordinary Petri nets superimposed with time for each transition.

Definition: *A Timed Petri net is a 5-tuple TPN* = (P, T, A, m_0, D), *where:*

1. *PN* (TPN) = (P, T, A, m_0) *is a P/T Petri net and*
2. $D: T \rightarrow R^+$ is the *duration function*, a mapping of each transition into a positive rational number, meaning firing of each transition t_i now takes dt_i time units.

Bibliographical Remarks
For an introduction to Petri nets, Reisig (2013) is recommended. Peterson (1981) is the first textbook on Petri nets, and it is still useful. For a summary of Petri nets, Murrata (1989) is recommended.

References

Cameron, A., Stumptner, M., Nandagopal, N., Mayer, W., & Mansell, T. (2015). Rule-based peer-to-peer framework for decentralised real-time service oriented architectures. *Science of Computer Programming, 97*, 202–234.

Cassandras, C. G., & Lafortune, S. (2009). *Introduction to discrete event systems*. Berlin: Springer Science & Business Media.

Davidrajuh, R., & Saadallah, N. (2016). Implementation of "Cohesive Place-Transition nets with Inhibitor Arcs" in GPenSIM. In *IEEE 2016 Asia Multi Conference on Modelling and Simulation*. Kota Kinabalu, Malaysia, December 4–6, 2016.

DiCesare, F., Harhalakis, G., Proth, J. M., Silva, M., & Vernadat, F. B. (1993). *Practice of Petri nets in manufacturing* (p. 8). London: Chapman & Hall.

Lopez, F., Barton, K., & Tilbury, D. (2017). *Simulation of discrete manufacturing systems with attributed hybrid dynamical nets* (unpublished in October 2017).

Murata, T. (1989). Petri nets: Properties, analysis and applications. *Proceedings of the IEEE, 77* (4), 541–580.

Peterson, J. L. (1977). Petri nets. *ACM Computing Surveys (CSUR), 9*(3), 223–252.

Peterson, J. L. (1981). *Petri net theory and the modeling of systems*. New Jersey, USA: Prentice-Hall.

Popova-Zeugmann, L. (2013). Time Petri nets. In *Time and Petri nets* (pp. 31–137). Berlin: Springer.

Reisig, W. (2013). *Understanding Petri nets: Modeling techniques, analysis methods, case studies*. Berlin: Springer.

Chapter 2
Modeling with GPenSIM: Basic Concepts

This chapter introduces the basic aspects of modeling discrete-event systems with GPenSIM. In GPenSIM, a clear separation of the static and the dynamic details is practiced. The static details of a Petri net are given in the Petri net Definition File (PDF), whereas the Main Simulation File (MSF) and the processor files contain the dynamic information of the Petri net.

2.1 Separating the Static and Dynamic Details

GPenSIM advocates a clear separation of the static and the dynamic details. The definition of a Petri net graph (*static* details) is given in the **Petri net Definition File** (**PDF**). There may be a number of PDFs if the Petri net model is divided into many modules, and each module is defined in a separate PDF. While the Petri net Definition File has the static details, the **Main Simulation File** (**MSF**) and the processor files contain the dynamic information (such as initial tokens in places, firing times of transitions) of the Petri net (Fig. 2.1).

2.2 Pre-processors and Post-processors

In addition to these two files (Main Simulation File—MSF and Petri net Definition File—PDF), there can be a number of pre-processors and post-processors. These processors define the run-time dynamic details of the model.

A pre-processor file contains the code for additional conditions to check whether an enabled transition can actually fire; in other words, pre-processor is run before firing a transition, just to make sure that an enabled transition can actually start

© The Author(s) 2018
R. Davidrajuh, *Modeling Discrete-Event Systems with GPenSIM*, SpringerBriefs in Applied Sciences and Technology, https://doi.org/10.1007/978-3-319-73102-5_2

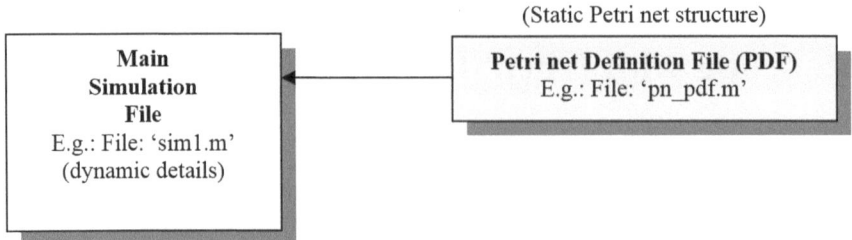

Fig. 2.1 Separating the static and the dynamic Petri net details

firing depending upon some other additional conditions ('firing conditions'). Further, we can write separate pre-processors for each transition or combine them into a single common pre-processor. It is also allowed to use individual pre-processors together with the common pre-processor.

A Post-processor file is run after the firing of a transition. A post-processor contains code for actions that has to be carried out after a certain transition completes firing. Just like pre-processors, post-processors can be specifically for individual transitions or combined into one common post-processor.

2.2.1 Using Pre-processor and Post-processor

According to the Petri net theory, a transition can fire ('enabled transition') when there are enough tokens in the input places. However, an event representing a transition can have additional restrictions for firing; for example, if two events (event-1 and event-2) are competing for a common resource and if they both are enabled, then event-2 is *allowed to fire* if event-2 possesses higher priority than event-1. In the GPenSIM literature, these additional conditions are called 'firing conditions.'

The firing conditions for firing a transition are coded in a pre-processor file. *After a transition completes firing*, there may be some bookkeepings need to be done or some other activities need to be performed; these can be coded into a post-processor file.

CAUTION! **Names of the processor files must follow a strict naming policy, as they will be chosen and run automatically. For example, the specific pre-processor for the transition 'trans1' must be named 'trans1_pre.m'; similarly, the specific post-processor for the transition 'trans1' must be named 'trans1_post.m'.**

2.3 Global Info

The different files (Main Simulation File MSF, Petri net Definition Files PDFs, and pre-processor and post-processor files) can access and exchange global parameter values through a packet called '**global_info**'. If a set of parameters is needed to be passed between different files, then these parameters are added to the **global_info** packet. Since **global_info** packet is visible in all the files, the values of the parameters in the packet can be read and even changed in different files.

2.4 Integrating with MATLAB Environment

One of the most important reasons for developing GPenSIM and the most advantage of it is its integration with the MATLAB environment so that we can harness diverse toolboxes available in the MATLAB environment; see Fig. 2.2. For example, by combining GPenSIM with the Control System Toolbox, we can experiment hybrid discrete–continuous control applications.

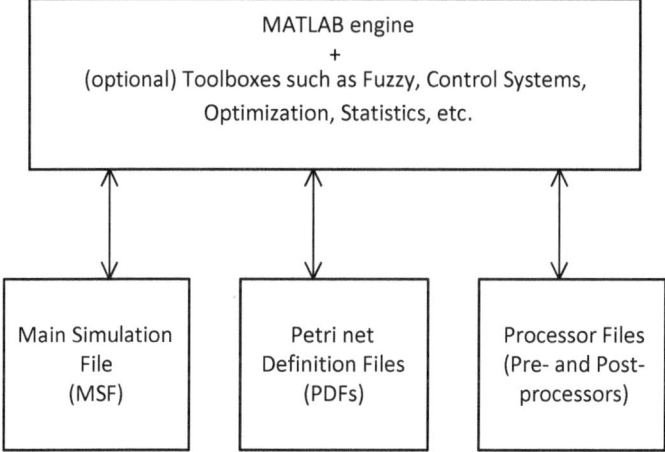

Fig. 2.2 Integrating with MATLAB environment

2.5   Creating an Untimed Petri net Model with GPenSIM

The methodology for creating a Petri net model consists of the following three
steps:

Step-1. Defining the Petri net graph in a Petri net Definition File (PDF): This is the
static part. This step consists of three substeps:

a. Identifying the passive elements of a Petri net graph: the places,
b. Identifying the active elements of a Petri net graph: the transitions, and
c. Connecting these elements with arcs.

Step-2. Coding the firing conditions in the relevant pre-processor files and
post-firing activities in the post-processor files
Step-3. Assigning the initial dynamics of a Petri net in the Main Simulation File
(MSF):

a. The initial markings on the places, and possibly
b. The firing times of the transitions.

After creating a Petri net model, simulations can be done.

2.5.1   Example-01: A Simple P/T (Untimed) Petri net

As the first example, we will create a P/T (untimed) Petri net. The three steps are
explained below, using the sample Petri net shown in Fig. 2.3.

2.5.2   Step-1: Defining the Petri net Graph

Defining the elements of a static Petri net graph is done in a Petri net Definition File
(PDF). PDF is to identify the elements (places, transitions) of a Petri net, and to
define the way these elements are connected. The Petri net graph shown in Fig. 2.3
has three places, one transition, and three arcs. The PDF for the graph is given
below:

Fig. 2.3 A simple Petri net

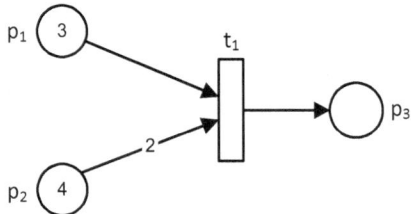

```
% Example-01: A Simple P/T Petri net
function [pns] = simple_pn_pdf()

pns.PN_name = 'A Simple P/T Petri net'; % module name or label
pns.set_of_Ps = {'p1', 'p2', 'p3'};  % set of places
pns.set_of_Ts = {'t1'};    % set of transitions
pns.set_of_As = {'p1','t1',1, 'p2','t1',2, 't1','p3',1}; % set of arcs
```

Explanation:

First, assign a name (or label) for the Petri net.
```
> PN_name = 'A Simple P/T Petri net';
```
Second, the places are to be identified with place names.
```
> set_of_Ps = {'p1', 'p2', 'p3'};
```

Third, the transitions are to be identified by stating their names.
```
> set_of_Ts = {'t1'};
```

Finally, how the elements are connected is defined: Connecting arcs are to be defined by listing the source, the destination, and the weights of each arc. For example, the first arc is from '**p1**' (source) to '**t1**' (destination) with a unit arc weight:

```
> set_of_As =  {'p1','t1',1, 'p2','t1',2, 't1','p3',1};
```

2.5.3 Step-2: Creating the Pre-processor and Post-processor Files

This example is simple in the sense the transition will always fire if enabled. Thus, there are no additional firing conditions to be coded in the pre-processor file. In addition, there is no need for post-processor file either, as no post-firing activity is specified.

2.5.4 Step-3: The Main Simulation File: Assigning the Initial Dynamics

After creating the PDF file (e.g., 'simple_pn_pdf.m'), we need to create the Main Simulation File (MSF). Sample MSF is given below:

```
% Example-01: A Simple P/T Petri net
% the Main Simulation File (MSF) to run simulation
global global_info
global_info.MAX_LOOP = 15; % limit simulation loops to 15

pns = pnstruct('simple_pn_pdf'); % create petri net structure

dyn.m0 = {'p1',3, 'p2',4};
pni = initialdynamics(pns, dyn);

Sim_Results = gpensim(pni); % perform simulation runs
prnss(Sim_Results); % print the simulation results
plotp(Sim_Results, {'p1','p2','p3'}); % plot the results
```

In the MSF, first we indicate the static Petri net graph, by passing the name of the PDF (**without** the ending '.m') to the function '**pnstruct**':

```
> pns = pnstruct('simple_pn_pdf'); % indicate the PDF
```

Second, the *initial dynamics* such as initial markings on the places are to be assigned. Normally, we stuff this information into a packet (e.g., 'dyn' in this example) and then pass this packet to the function 'initialdynamics'. In the code given below, there are three and four initial tokens in the places **p1** and **p2**, respectively. The function 'initialdynamics' creates the initial Petri net dynamic structure.

```
> dyn.m0 = {'p1',3, 'p2',4};    % the intial markings
> pni = initialdynamics(pns, dyn); % combine static & intial dynamics
```

2.5.5   The Simulations

Function **gpensim** will do the simulations if the initial Petri net dynamic structure is passed to it:

```
> Sim_Results = gpensim(pni); % run the simulations
```

The output parameter of gpensim 'Sim_Results' is the simulation results. Sim_Results is a structure for the simulation results. To view the simulation results, the function '**prnss**' (meaning 'print state space') could be used.

2.5.6   Viewing the Simulation Results with 'Prnss'

```
> prnss(Sim_Results);
```

The output is given below:

```
======= State Diagram =======
**     Time: 0    **
State:0 (Initial State): 3p1 + 4p2
At start ....
At time: 0,  Enabled transitions are:    t1
At time: 0,  Firing transitions are:     t1

**     Time: 0    **
State: 1
Fired Transition: t1
Current State: 2p1 + 2p2 + p3
Virtual tokens: (no tokens)

Right after new state-1 ....
At time: 0,  Enabled transitions are:    t1
At time: 0,  Firing transitions are:     t1


**     Time: 0    **
State: 2
Fired Transition: t1
Current State: p1 + 2p3
Virtual tokens: (no tokens)

Right after new state-2 ....
At time: 0,  Enabled transitions are:
At time: 0,  Firing transitions are:
```

In addition to the text printout, we can also view the results graphically. For example,

```
> plotp(Sim_Results, {'p1', 'p2', 'p3'});
```

The above statement will plot how the tokens in the places vary with time: See Fig. 2.4 given below:

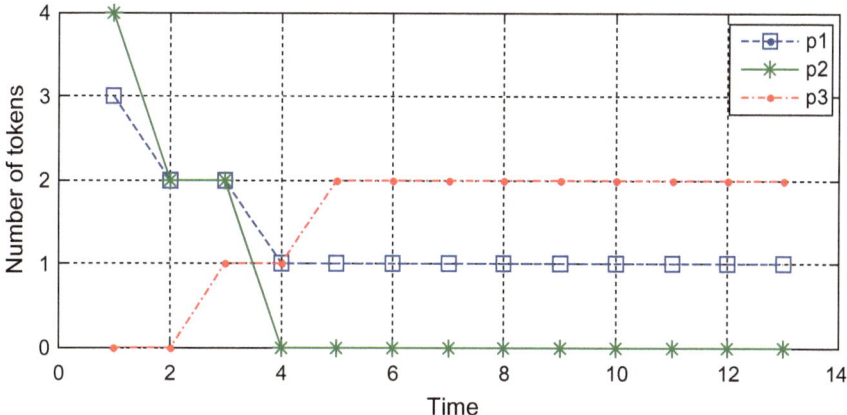

Fig. 2.4 Simulation results of example-01

2.6   Summary: Creating a Simple Untimed Petri net Model

Step-1: This step is about creating the PDF that defines the static Petri net graph.
The PDF for the Petri net shown in Fig. 2.3 is repeated below:

```
% Example-01: A Simple P/T Petri net
% PDF: 'simple_pn_pdf.m'
function [pns] = simple_pn_pdf()

pns.PN_name = 'A Simple P/T Petri net';
pns.set_of_Ps = {'p1', 'p2', 'p3'};
pns.set_of_Ts = {'t1'};
pns.set_of_As = {'p1','t1',1, 'p2','t1',2, 't1','p3',1};
```

Step-2: For the example-01, step-2 is not needed as there is no need for
pre-processor and post-processor files.
Step-3: This step is for assigning the initial dynamics (initial markings, firing times,
etc.) in the MSF. After the assignment, the simulations can be run and the results
can also be plotted. The MSF for the Petri net shown in Fig. 2.3 is repeated below:

```
% Example-01: A Simple P/T Petri net
% the Main Simulation File (MSF) to run simulation
global global_info
global_info.MAX_LOOP = 15; % limit simulation cycles to 15
pns = pnstruct('simple_pn_pdf'); % create petri net structure
dyn.m0 = {'p1',3, 'p2',4};
pni = initialdynamics(pns, dyn);

Sim_Results = gpensim(pni); % perform simulation runs
prnss(Sim_Results); % print the simulation results
plotp(Sim_Results, {'p1','p2','p3'}); % plot the results
```

2.7   Static PN Structure

In the Main Simulation File given in the previous subsection, first we get a *static*
Petri net structure (called **pns** in the code shown above) as the output parameter of
function **pnstruct**:

```
pns = pnstruct('simple_pn_pdf');
```

The static PN structure **pns** is a compact representation of the static Petri net graph. A static PN structure consists of five elements, e.g., in **pns**:

```
                name: 'A Simple Petri net'
       global_places: [1x3 struct]
         No_of_places: 3
  global_transitions: [1x1 struct]
   No_of_transitions: 1
      global_Vplaces: [1x3 struct]
     incidence_matrix: [1 2 0 0 0 1]
```

The elements of a static PN structure are:

(1) name: the text string identifier of the Petri net,
(2) global_places: the set of all places in the Petri net,
(3) global_transitions: the set of all transitions in the Petri net,
(4) global_Vplaces: the tokens that are residing inside firing transitions, and
(5) incidence_matrix: the matrix that depicts how the places and transitions are connected together.

It must be emphasized that *static* PN structure is much simpler than *run-time* PN structure. During simulation ('run-time'), state information and the other run-time information will be added to the PN structure; thus, the PN structure will contain dynamic information in addition to static details; during simulation, the PN structure is called 'run-time' PN structure. Details of run-time PN structure are given in Chap. 8.

2.8   Assigning Names to Places and Transitions

CAUTION! There is a serious restriction in naming: ONLY the first 10 characters of NAMES are significant.

This means names for two places (**pReggieDavidrajuh_1**) and (**pReggieDavidrajuh_2**) are the same names (REFER TO THE SAME PLACE) because the first 10 characters (**pReggieDav**) of these two names are the same.

However, (**pReggie_1_Davidrajuh**) and (**pReggie_2_Davidrajuh**) are different names simply because the first 10 characters of these two names are different in this case.

2.9   GPenSIM Reserved Words

CAUTION! There are a few reserved words in GPenSIM. Using these words as variable names should be avoided.

Table 2.1 shows some words that are reserved for GPenSIM internal usage. Avoid using these reserved words as the names of your variables.

2.10   Creating a Timed Petri net

As the second example, we will create a simple **Timed** Petri net; this example is almost the same as the previous example-01 (Fig. 2.3) except the fact that the **transition(s) are assigned firing time to make a Timed Petri net.**

2.10.1   Example-02: A Simple Timed Petri net

PDF: PDF will be the same as before as there is no change in the static Petri net graph (Fig. 2.3):

```
% Example-02: A Simple Timed Petri net
% PDF: 'simple_tpn_pdf.m'
function [pns] = simple_tpn_pdf()

pns.PN_name = 'A Simple Timed Petri net '; % name or label
pns.set_of_Ps = {'p1', 'p2', 'p3'};    % set of places
pns.set_of_Ts = {'t1'};   % set of transitions
pns.set_of_As = {'p1','t1',1, 'p2','t1',2, 't1','p3',1};%set of arcs
```

Table 2.1 Reserved words

Reserved word	Meaning
PN	During simulations, **PN** is a large structure that represents the whole Petri net dynamics. PN is visible as a global variable in all GPenSIM system files
global_info	**global_info** is also a global variable; global_info carries user-defined variables to all the system files. **global_info** also stores the global OPTIONS; OPTIONS are discussed in Chap. 5

MSF:

For the Timed Petri net, we add firing times to all the transitions.

MSF for example-02:

```
% Example-02: A Simple Timed Petri net
% the Main Simulation File (MSF) to run simulation
clear all; clc;
global global_info % user data
global_info.STOP_AT = 50;     % stop at time = 50 TU

pns = pnstruct('simple_tpn_pdf'); % create petri net structure

dyn.m0 = {'p1',3, 'p2',4};  % initial tokens
dyn.ft = {'t1',10};   % firing times
pni = initialdynamics(pns, dyn);
sim = gpensim(pni); % perform simulation runs
prnss(sim); % print the simulation results
plotp(sim, {'p1','p2','p3'}); % plot the results
```

In the MSF, there are two changes from the previous example:

The first change is to instruct the simulator to stop after 50 time units as this is a Timed Petri net.

```
> global_info.STOP_AT = 50;     % stop at time = 50 TU
```

The second change is to assign firing time to the only transition **t1**.

```
> dyn.ft = {'t1', 10};     % firing time of 't1' is 10 TU
```

The Simulations:

Function 'gpensim' will do the simulations if the Petri net marked graph is passed to it. The results can be echoed (text printout) on the screen by 'prnss'; in addition to the text printout, we can also view the results graphically using the function 'plotp'.

```
> Sim_Results = gpensim(pni);
> prnss(Sim_Results);
> plotp(Sim_Results, {'p1','p2','p3'});
```

The output is given below:

Simulation Results:

Of course, **t1** takes 10 TU (e.g., milliseconds) to produce a token on **p3**, after removing 1 and 2 tokens from **p1** and **p2**, respectively.

```
======= State Diagram =======
**     Time: 0    **
State:0 (Initial State): 3p1 + 4p2
At start ....
At time: 0,  Enabled transitions are:      t1
At time: 0,  Firing transitions are:       t1


**     Time: 10    **
State: 1
Fired Transition: t1
Current State: 2p1 + 2p2 + p3
Virtual tokens: (no tokens)

Right after new state-1 ....
At time: 10,  Enabled transitions are:      t1
At time: 10,  Firing transitions are:       t1


**     Time: 20    **
State: 2
Fired Transition: t1
Current State: p1 + 2p3
Virtual tokens: (no tokens)

Right after new state-2 ....
At time: 20,  Enabled transitions are:
At time: 20,  Firing transitions are:
```

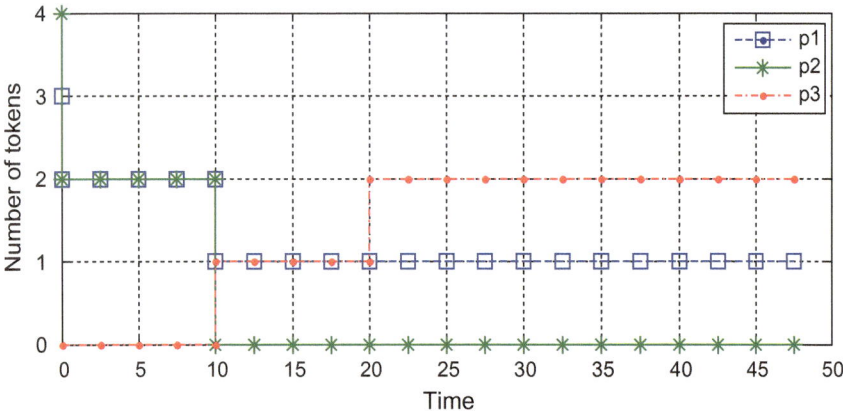

Fig. 2.5 Simulation results of example-02

The function **plotp** plots how the tokens in the places vary with time, as shown in Fig. 2.5.

Bibliographical Remarks
One of the main design goals of GPenSIM is to let Petri net models be combined with the other toolboxes that are available on the MATLAB platform. Many useful solutions were obtained by combining Petri nets with Fuzzy Logic (e.g., Pedrycz

and Gomide 1994), feedback control (e.g., Yamalidou et al. 1996), linear programming (e.g., Silva et al. 1998), and discrete optimization (e.g., Nishi and Maeno 2010).

References

Nishi, T., & Maeno, R. (2010) Petri net decomposition approach to optimization of route planning problems for AGV systems. *IEEE Transactions on Automation Science and Engineering, 7*(3), 523–537.

Pedrycz, W., & Gomide, F. (1994). A generalized fuzzy Petri net model. *IEEE Transactions on Fuzzy Systems, 2*(4), 295–301.

Silva, M., Terue, E., & Colom, J. M. (1998). Linear algebraic and linear programming techniques for the analysis of place/transition net systems. In *Lectures on Petri nets I: Basic Models* (pp. 309–373). Berlin: Springer.

Yamalidou, K., Moody, J., Lemmon, M., & Antsaklis, P. (1996). Feedback control of Petri nets based on place invariants. *Automatica, 32*(1), 15–28.

Chapter 3
Pre-processor and Post-processor Files

This chapter introduces the processor files of the GPenSIM. With the processors, many real-life discrete event systems could be modeled as Petri nets. The processors also give GPenSIM the flexibility it is well known for and the potential for making hybrid Petri net models by combining Petri nets with the other MATLAB Toolboxes (e.g., Fuzzy Logic, Control Systems).

As discussed in the previous chapter, the methodology for creating a Petri net model consists of three steps. These steps are (1) defining the Petri net graph in a PDF, (2) coding the firing conditions in the relevant pre-processor files and post-firing activities in the post-processor files, and (3) assigning the initial dynamics of a Petri net in the MSF. However, in the previous chapter, step-2 was omitted as there were no additional firing conditions to be coded (in the pre-processor files) and there were no post-firing activities (to be coded in the post-processor files). In this chapter, the processor files (pre- and post-processor files) are studied.

3.1 The Processor Files

There are four types of processor files:

- **Specific pre-processor file**:
  This file is run **before** firing a specific transition. This file is coded with firing conditions for a specific transition. Thus, when a transition is enabled, the corresponding pre-processor file, if it exists, will be checked to make sure all the firing conditions coded in this file are met. Only if all the firing conditions coded in the pre-processor file are satisfied (this file returns a singleton value '1'), then the transition can start firing.

© The Author(s) 2018
R. Davidrajuh, *Modeling Discrete-Event Systems with GPenSIM*, SpringerBriefs in Applied Sciences and Technology, https://doi.org/10.1007/978-3-319-73102-5_3

Let us say that a transition **t1** is enabled. If the specific pre-processor file of this transition 't1_pre' exists then it will be run. Only if 't1_pre' returns a singleton (logic value '1'), then the enabled **t1** is allowed to start firing immediately.

- **Specific post-processor file**:
  This file is coded with post-firing work that may be necessary **after** firing of a specific transition.
  Let us say that **t1** fires; right after the firing, if the file '**t1_post**' exists then it will also be run.

- **COMMON_PRE** file:
  Just like the specific pre-processor files that are specific to individual transitions, COMMON_PRE file, if exists, will be run **before** firing of **every** transition. Thus, COMM_PRE is **common** (one and only one for all transitions) and contains firing conditions that all the enabled transitions must satisfy to start firing.
  Let us consider again the situation where **t1** is enabled. If the specific pre-processor file **t1_pre** exists then it will be run. Also, if the file **COMMON_PRE** exists, it will also be run. Only if **both** **t1_pre** and COMMON_PRE return logic 1 value, then the enabled **t1** is allowed to fire.

- **COMMON_POST** file:
  This file is coded with post-firing activities or accounting work needed to be done after firing of every transition. Let us say that **t1** fires; right after the firing completes, if the file **t1_post** exists then it will be run. Also, if the **COMMON_POST** file exists then it will also be run.

3.2 Structure of a Pre-processor File

Given below is the skeleton of a specific pre-processor file:

```
function [fire, transition] = tRobot_1_pre(transition)
% function [fire, transition] = tRobot_1_pre(transition)
…
…
fire = ;
```

The **input parameter is**:

(1) **transition**: Input parameter 'transition' is a structure representing the enabled transition. This structure has many fields including 'transition.name' which is the name of the enabled transition.

The **output parameters** are:

(1) **fire: must be logic 'true' for firing**; if the returned value for fire is logic false, then the enabled transition cannot fire; the enabled transition will be blocked.

(2) **transition**: A structure representing the enabled transition that will carry the following information back to the calling function:

   a. **transition.new_color**: (coloring of tokens is not discussed in this book),
   b. **transition.override**: (coloring of tokens is not discussed in this book),
   c. **transition.selected_tokens**: (coloring of tokens is not discussed in this book).

3.2.1   Example-03: Pre-processor Example

Figure 3.1 shows a Petri net model of a production facility where three robots are involved in sorting products from an input buffer to output buffers. The three robots are represented by the transitions **Robot-1** to **Robot-3** and the buffers by the places **Buffer-1** to **Buffer-3**. Let us assume that the three robots take 10, 5, and 15 time units per operations.

Let us add some **conditions for firing**:

1. The output buffers have limited capacity: Buffer-1, Buffer-2, and Buffer-3 can accommodate a maximum of 2, 3, and 1 machined parts (tokens), respectively.
2. The robots should be operated in a manner that, at any time, Buffer-2 should have more parts than Buffer-1 and Buffer-1 should have more parts than Buffer-3.

The firing conditions stated above shall be coded in the pre-processor files.

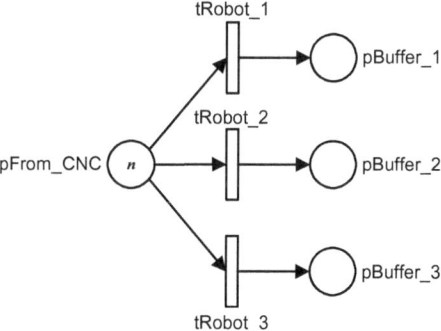

Fig. 3.1 Petri net model of a production facility

Creating M-Files:

In this example, the following M-files are created in the three steps:

- Step-1: Creating the **PDF** file.
- Step-2: Creating the **pre-processors for the three transitions**. This is because there are firing conditions attached to the transitions.
- Step-3: Creating the **MSF**: Assigning the initial dynamics (initial markings and firing times) and running the simulations.

Step-1: PDF: Defining the Petri net graph
Let us call the PDF for the Petri net in Fig. 3.1 as 'pre_processor_pdf.m':

```
% Example-03: Pre-processor Example
% file: pre_processor_pdf.m: definition of petri net
function [pns] = pre_processor_pdf()

pns.PN_name = 'Pre-processor Example: PN for production facility';

pns.set_of_Ps = {'pFrom_CNC', 'pBuffer_1', 'pBuffer_2', 'pBuffer_3'};
pns.set_of_Ts = {'tRobot_1','tRobot_2','tRobot_3'};
pns.set_of_As = {'pFrom_CNC','tRobot_1',1, 'pFrom_CNC','tRobot_2',1, ...
    'pFrom_CNC','tRobot_3',1,...
    'tRobot_1','pBuffer_1',1, 'tRobot_2','pBuffer_2',1,...
    'tRobot_3','pBuffer_3',1};
```

Step-2: Pre-processor files:
We will need three pre-processor files, one for each transition:

- Pre-processor **'tRobot_1_pre'**: tRobot_1 will fire only if the number of tokens (machined parts) already put in output **pBuffer_1** is less than 2. In addition, number of tokens in **pBuffer_1** should be less than that of **pBuffer_2**; coding these two firing conditions into the pre-processor for **tRobot-1** is given below. As the name of the transition is '**tRobot_1**', this pre-processor must be named '**tRobot_1_pre.m**'.

```
function [fire, transition] = tRobot_1_pre(transition)

n1 = ntokens('pBuffer_1');   % check how many tokens in pBuffer-1
n2 = ntokens('pBuffer_2');   % check how many tokens in pBuffer-2
fire=and(lt(n1,n2),lt(n1,2)); % Robot_1 can fire ONLY if these 2 satisfied
```

- Similarly, the pre-processor files for **tRobot_2** and **tRobot_3** are created, satisfying the given firing conditions:

```
function [fire, transition] = tRobot_2_pre(transition)

n2 = ntokens('pBuffer_2');   % check how many tokens in pBuffer-2
fire = lt(n2, 3);   % tRobot_2 can fire ONLY if n2 is less than 3
```

```
function [fire, transition] = tRobot_3_pre(transition)

n1 = ntokens('pBuffer_1');    % check how many tokens in pBuffer-1
n3 = ntokens('pBuffer_3');    % check how many tokens in pBuffer-3
fire =and(gt(n1,n3),lt(n3,1)); % Robot_3 can fire ONLY if these 2 satisfied
```

Step-3: MSF: Assigning the initial dynamics and running simulations
Given below is the Main Simulation File ('pre_processor.m'):

```
% Example-03: Pre-processor Example
global global_info
global_info.STOP_AT = 60; % stop after 60 time units

pns = pnstruct('pre_processor_pdf');
dyn.m0 = {'pFrom_CNC',8}; % tokens initially
dyn.ft = {'tRobot_1',10, 'tRobot_2',5, 'tRobot_3',15};  % firing times
pni = initialdynamics(pns, dyn);

sim = gpensim(pni);
prnss(sim);
plotp(sim, {'pBuffer_1', 'pBuffer_2', 'pBuffer_3'});
```

The printout of **prnss** given below is one of the two possible outcomes.

Outcome-1:

```
     Time: 0
State:0 (Initial State)
pBuffer_1 pBuffer_2 pBuffer_3 pFrom_CNC
 0        0         0         8
At time: 0
  Enabled transitions are:
 tRobot_1    tRobot_2    tRobot_3
At time: 0
  Firing transitions are:
 tRobot_2

     Time: 5
State: 1
Fired Transition: tRobot_2
Current State:
pBuffer_1 pBuffer_2 pBuffer_3 pFrom_CNC
 0        1         0         7
At time: 5
  Enabled transitions are:
 tRobot_1    tRobot_2    tRobot_3
At time: 5
  Firing transitions are:
 tRobot_1    tRobot_2
```

```
      Time: 10
State: 2
Fired Transition: tRobot_2
Current State:
pBuffer_1 pBuffer_2 pBuffer_3 pFrom_CNC
  0         2         0         5
At time: 10
   Enabled transitions are:
  tRobot_1    tRobot_2    tRobot_3
At time: 10
   Firing transitions are:
  tRobot_1    tRobot_2

      Time: 15
State: 3
Fired Transition: tRobot_2
Current State:
pBuffer_1 pBuffer_2 pBuffer_3 pFrom_CNC
  0         3         0         4
At time: 15
   Enabled transitions are:
  tRobot_1    tRobot_2    tRobot_3
At time: 15
   Firing transitions are:
  tRobot_1

      Time: 15
State: 4
Fired Transition: tRobot_1
Current State:
pBuffer_1 pBuffer_2 pBuffer_3 pFrom_CNC
  1         3         0         4
At time: 15
   Enabled transitions are:
  tRobot_1    tRobot_2    tRobot_3
At time: 15
   Firing transitions are:
  tRobot_1    tRobot_3

      Time: 25
State: 5
Fired Transition: tRobot_1
Current State:
pBuffer_1 pBuffer_2 pBuffer_3 pFrom_CNC
  2         3         0         2
At time: 25
   Enabled transitions are:
  tRobot_1    tRobot_2    tRobot_3
At time: 25
   Firing transitions are:
  tRobot_3

      Time: 30
State: 6
Fired Transition: tRobot_3
Current State:
pBuffer_1 pBuffer_2 pBuffer_3 pFrom_CNC
  2         3         1         2
At time: 30
   Enabled transitions are:
  tRobot_1    tRobot_2    tRobot_3
At time: 31.25
   Enabled transitions are:
  tRobot_1    tRobot_2    tRobot_3
>>
```

Figure 3.2 shows the plot of how the number of tokens in different places varies with time:

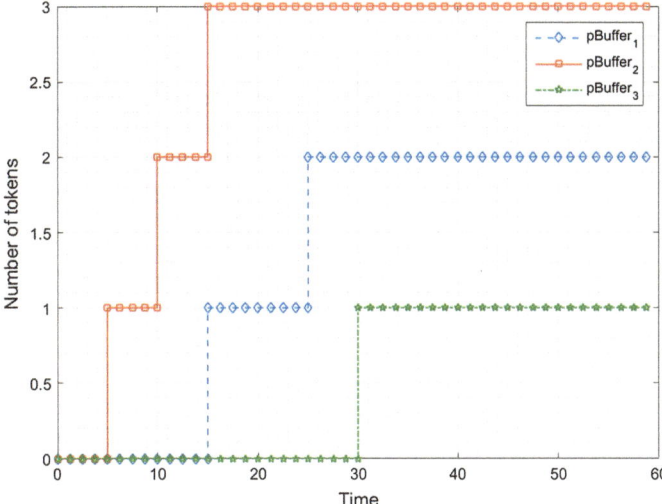

Fig. 3.2 Simulation results

3.2.2   Example-04: COMMON_PRE Example

In the previous example, we used three pre-processor files (namely **tRobot_1_pre**, **tRobot_2_pre**, and **tRobot_3_pre**) to check whether the enabled transitions can fire. If we had ten robots, then we have to create ten pre-processor files—this may become tiresome. Instead of individual (specific) pre-processor files, we can create one and only one common COMMON_PRE file. The section on '**COMMON PROCESSORS**' presents the details. For example, COMMON_PRE file that is given below will **replace** the three individual pre-processor files.

```
function [fire, transition] = COMMON_PRE(transition)
% function [fire, trans] = COMMON_PRE(trans)

n1 = ntokens('pBuffer_1');   % check how many tokens in pBuffer-1
n2 = ntokens('pBuffer_2');   % check how many tokens in pBuffer-2
n3 = ntokens('pBuffer_3');   % check how many tokens in pBuffer-3

% checking firing conditions for tRobot_1
if (strcmp(transition.name, 'tRobot_1')),
    fire = and(lt(n1, n2), lt(n1, 2);

% checking firing conditions for tRobot_2
elseif (strcmp(transition.name, 'tRobot_2')),
    fire = lt(n2, 3);

% checking firing conditions for tRobot_3
elseif (strcmp(transition.name, 'tRobot_3')),
    fire = and(gt(n1, n3), lt(n3, 1);

else % this is not possible
    error('transition name is neither of the three robots ...')
end
```

3.3   Implementing Preference Through Pre-processors

When more than one enabled transition competes for a common resource (e.g.,
common input tokens), sometimes it is better to have a preference order so that the
firing is deterministic rather than allowing an arbitrary transition to fire. In such a
situation, pre-processor and COMMON_PRE can be used to block some transitions
so that some others (with higher priority) are allowed to fire. A better way to do this
is to assign priorities to the transitions so that GPenSIM automatically chooses (no
coding necessary) the transition with the highest priority when there is competition;
the topic of transitions with priority is discussed in Sect. 6.5. However, in this
section, we will see how we can achieve preference through pre-processor/
COMMON_PRE.

3.3.1   Example-05: Implementing Preference Through
Pre-processors

In this example (Fig. 3.3), transitions **t1** and **t2** both compete for tokens in **pS**; let us
say that we will allow **t1** to fire and **t2** to wait. The conditions that allowing **t1** to
fire while blocking **t2** are coded in the pre-processor files.
   MSF:

```
% MSF: Example-05: prefer.m
global global_info
global_info.STOP_AT = 60;    % Stop after 60 time units

% CASE = 1: t1 is preferred;
% CASE = 2: t2 is preferred
%    otherwise, no preference
global_info.CASE = 1;

pns = pnstruct('prefer_pdf');

dyn.ft = {'allothers',10};  % firing times for all the transition is 10 TU
dyn.m0 = {'pS',4};
pni = initialdynamics(pns, dyn);
sim_results = gpensim(pni);
plotp(sim_results, {'pE1', 'pE2'});
```

Fig. 3.3 Petri net with two
competing transitions

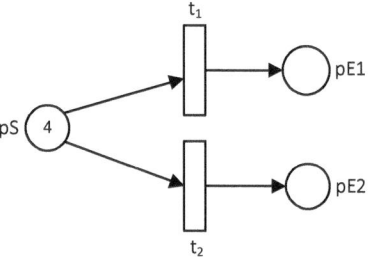

PDF:

```
% Example-05: prefer_pdf.m
function [pns] = prefer_pdf()
pns.PN_name = 'Preference example';
pns.set_of_Ps = {'pS', 'pE1', 'pE2'};
pns.set_of_Ts = {'t1','t2'};
pns.set_of_As = {'pS','t1',1, 't1','pE1',1, 'pS','t2',1, 't2','pE2',1};
```

Pre-processor t1_pre:

```
function [fire, transition] = t1_pre (transition)
global global_info
if eq(global_info.CASE, 2),
    % Case = 2: t2 is preferred
    % thus, if t2 is also enabled, then block t1
    fire = not(is_enabled('t2'));
else
    % Case-1: t1 is preferred
    % thus, fire t1, whenever it is enabled
    fire = 1;
end
```

Pre-processor: t2_pre:

```
function [fire, transition] = t2_pre (transition)
global global_info
if eq(global_info.CASE, 1),
    % Case = 1: t1 is preferred,
    % In this case, block t2, if t1 is enabled
    fire = not(is_enabled('t1'));
else
    % Case-2: t2 is preferred,
    % thus, fire t2, if it is enabled
    fire = 1;
end
```

Simulation results:
The simulation result (Fig. 3.4) shows that for case-1, **t1** is preferred and thus it is allowed to fire all four times, filling its output place **pE1** with four tokens.

In case-2 (Fig. 3.5), **t2** is allowed to fire all four times, filling its output place **pE2** with four tokens.

3.4 Post-processors

As stated in the earlier sections, there are two types of processors:

- **Pre-processor**, which is executed *before* firing a transition, just to check whether an enabled transition can start firing, by checking all the firing conditions given in the pre-processor file.
- **Post-processor**, which is executed *after* firing of a transition, is to perform any post-firing activities.

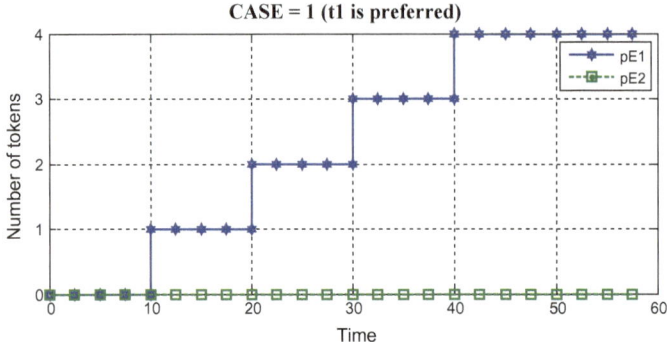

Fig. 3.4   Simulation results

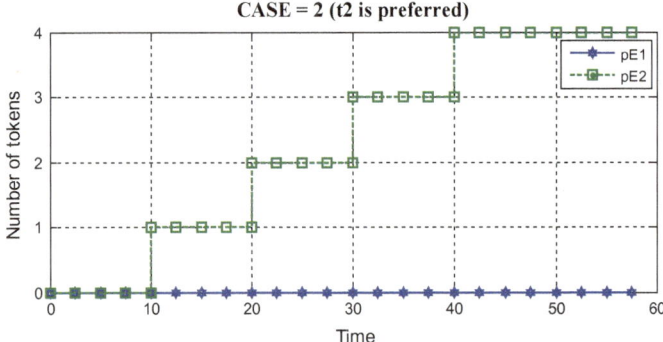

Fig. 3.5   Simulation results

The skeleton of a post-processor file is given below:

```
function [] = tX1_post(transition)
% function tX1_post
...
```

The **input parameter: transition**
transition is a structure that contains the **name of the fired transition**
There are no **output parameters**.

Note: In the post-processor file,

- If we are going to use any variables in the **global_info** packet, then we must declare **global_info** as a global variable.
- We can also access the run-time **Petri net structure** (**PN**) by declaring **PN** as a global variable too.

3.4.1 Example-06: Alternating Firing Using Binary Semaphore

Figure 3.6 depicts a Web server consisting of two server machines that serve clients *alternatingly*. First, client requests are queued at **pSTART**. Then two routers (**tX1** and **tX2**) remove the client requests from the **pSTART** and put it to the queue **p1** for Web server-1 and the queue **p2** for Web server-2. To evenly distribute client requests to both servers, one would expect that the two routers **tX1** and **tX2** fire alternatingly, meaning that no router fires more times than the other.

To allow the routers (transitions **tX1** and **tX2**) fire alternatingly, we can implement a binary semaphore that can be read and manipulated by the specific processor files of both transitions.

PDF: ('loadbalance_pdf.m'):

```
% Example-06: Load balance with Binary semaphore
function [pns] = loadbalance_pdf()
pns.PN_name = 'Load Balancer with binary semaphore';
pns.set_of_Ps = {'pSTART', 'p1', 'p2'};
pns.set_of_Ts = {'tX1','tX2'};
pns.set_of_As = {'pSTART','tX1',1, 'tX1','p1',1,...
                 'pSTART','tX2',1, 'tX2','p2',1};
```

Main Simulation File ('loadbalance.m'):

```
% Example-06: Load balancing with Binary semaphore
global global_info
global_info.semafor = 1;    % GLOBAL DATA: binary semafor
global_info.STOP_AT = 100;  % Option: stop the simulations after 100 TU

pns = pnstruct('loadbalance_pdf');
dynamicpart.m0 = {'pSTART', 10};
dynamicpart.ft = {'tX1', 10, 'tX2', 20};
pni = initialdynamics(pns, dynamicpart);
sim = gpensim(pns);
plotp(sim, {'p1', 'p2'});
```

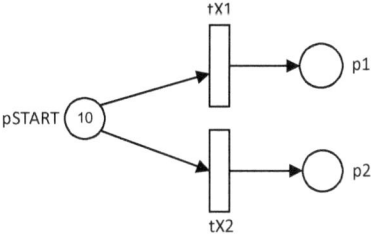

Fig. 3.6 Load balancing by alternating firing

Note: The parameter **semafor** with an initial value of 1 is added to the global_info packet. The initial value of '1' for semafor means the transition **tX1** is to fire first.

Pre-processor for tX1 ('tX1_pre.m'):

```
function [fire, transition] = tX1_pre(transition)
global global_info
% tX1 fires only if value of semaphore is 1
fire =   eq(global_info.semafor,1);
```

Post-processor for tX1 ('tX1_post.m'):

```
function [] = tX1_post(transition)
global global_info
% after firing tX1, set the semapfore to 2, so that tX2 can fire next
global_info.semafor = 2; % release semafor to tX2
```

Pre-processor for tX2 ('tX2_pre.m'):

```
function [fire, transition] = tX2_pre(transition)
global global_info
% tX2 fires only if value of semaphore is 2
fire = eq(global_info.semafor,2);
```

Post-processor for tX2 ('tX2_post.m'):

```
function [] = tX2_post(transition)
global global_info
% after firing tX2, set the semapfore to 1, so that tX1 can fire next
global_info.semafor = 1; % release semafor to tX1
```

The plot given below shows that the queues (**p1** and **p2**) are filled evenly. This is because of the transitions firing alternatingly.

3.5 Common Processors

In the examples we saw up to now, the processors are specific for individual transitions. This means, if there are n transitions, we may have to code a maximum of n pre-processor files and n post-processor files, which may make up too many processor files. In most cases, these files are almost equal, except for the name of the transitions.

GPenSIM allows common pre-processor and post-processor files for all the transitions, making it unnecessary to have too many specific processor files. If we code a flexible COMMON_PRE and COMMON_POST files, we need just four files (MSF, PDF, COMMON_PRE, and COMMON_POST) for simulation of a system.

Caution! Naming of the two files must be all capitalized: 'COMMON_PRE' and 'COMMON_POST'.

3.6   Structure of COMMON_PRE and COMMON_POST Files

The skeleton of the COMMON_PRE file:

```
function [fire, transition] = COMMON_PRE(transition)
...
...
fire =
```

The **input parameters** are:

(1) **transition: the structure that posses the <u>name</u> of the <u>enabled</u> transition (transition.name)**.

The **output parameters** are:

(1) **fire: must be logic true for firing**; if fire is equal to logic false, then the enabled transition cannot fire—it is blocked.
(2) **transition**: a structure to which we can add values like **new_color**, **override**, and **slected_tokens**.

Given below is the skeleton of the **COMMON_POST** file:

```
function [] = COMMON_POST(transition)
...
...
```

The **input parameter** is **transition: a structure that posses <u>name</u> of the <u>fired</u> transition**
The **output parameters** are <u>**NONE**</u>

3.6.1   Example-07: Alternating Firing with COMMON Processors

This example is exactly same as the example-06. However, in this example, we are going to use COMMON_PRE and COMMON_POST files for manipulating the binary semaphore.

- PDF 'loadbalance_pdf.m': Same as the one given in example-06.
- MSF 'loadbalance.m': Same as the one given in example-06.
- Specific pre-files 'tX1_pre.m' and tX2_pre.m' are obsolete; these two files are going to be replaced by the file 'COMMON_PRE.m', which is given below.

- Specific post-files 'tX1_post.m' and tX2_post.m' are obsolete; these two files are
  going to be replaced by the file 'COMMON_POST.m', which is given below.
- The simulation results are the same as the one shown in Fig. 3.7.

COMMON_PRE.m:

```
function [fire, trans] = COMMON_PRE(trans)
% COMMON_PRE file codes the enabling conditions

% Here, the current value of the semaphore
% indicate which transiton can fire.
global global_info
if strcmp(trans.name, 'tX1'),
    % for tX1: is semafor == 1?
    fire = eq(global_info.semafor, 1);

elseif strcmp(trans.name, 'tX2'),
    % for tX2: is semafor == 2?
    fire = eq(global_info.semafor, 2);

else % transition is neither "t1" nor "t2"? NOT possible
end
```

COMMON_POST.m:

```
function [] = COMMON_POST(transition)
% COMMON_POST file codes the post firing actions.

% Here, right after firing, the fired transition
% set the value of semaphore to the other
% transition so that the other one fires next
global global_info
if strcmp(transition.name, 'tX1'),
    % for tX1: set semafor to 2
    global_info.semafor = 2; % t1 releases semafor to t2

elseif strcmp(transition.name, 'tX2'), % transition.name 't2'
    % for tX2: set semafor to 1
    global_info.semafor = 1; % t2 releases semafor to t1
else   % This is not possible
end
```

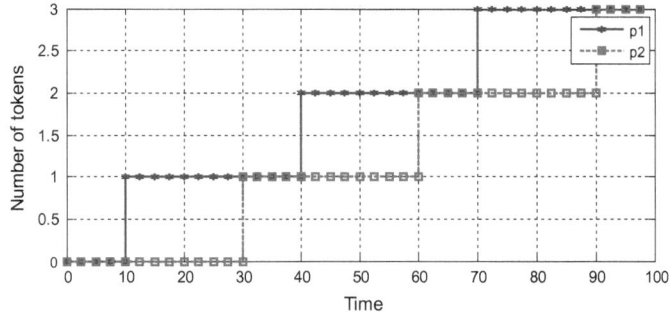

Fig. 3.7 Binary semaphore in action

3.7 Summary: COMMON Processors Versus Specific Processors

For modeling a Petri net with n transitions: In addition to the MSF and PDFs, we need the following processor files:

- *On one extreme*, we can code the system with a maximum of n specific pre-processor files and n specific post-processor files. In addition, we can have one COMMON_PRE and one COMMON_POST files as well; this makes $(2n + 2)$ processor files.
- *On the other extreme*, we can code the system with just two (2) processor files: COMMON_PRE and COMMON_POST files.

So, which is better? Perhaps, a combination of these two should be used; common actions can be put naturally in the common files, and actions that are specific and heavy (in terms of coding) can be put in the specific files. **In real-time environments, it is not a good idea to overload the common files; in a real-time environment, light common files supplemented by specific processor files are ideal** (Table 3.1).

3.7.1 Example-08: Mixing COMMON and Specific Processors

Let us consider a simple Petri net shown in Fig. 3.8. In this example, transition **tS** can fire at will ('cold firing'), as it is not restricted by any input places. Let us say that **tS** should fire exactly ten times. The other transitions **tA**, **tB**, and **tC**, should fire exactly three times each. This means, at the 'end of the day', there will be one token left at **pS**, and the other places **pA**, **pB**, and **pC** will have three tokens each.

Since **tA**, **tB**, and **tC** fire similarly (three times each), we can code this condition in the COMMON_PRE file. Since **tS** fires differently (ten times) than the other transitions, let us code this condition in the specific pre-processor file for **tS** 'tS_pre. m'. Of course, we could use one and only one COMMON_PRE to tackle the different number of firings, but the idea is to show that we can combine both COMMON_PRE and specific pre-processors.

Table 3.1 Processor files: COMMON versus specific

	COMMON PROCESSOR files	Specific processor files
Advantage	Fewer files (just two files: COMMON_PRE and COMMON_POST)	Easy to understand as it uses specific names and variables
Disadvantage	Hard to understand; very generic and can become large	Many files $(2n)$

Fig. 3.8 Using COMMON
and specific processors

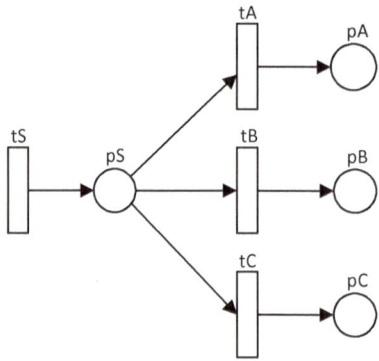

PDF:

```
% Example-08: Using COMMON and specific pre files together
% file: cas_pdf.m: PDF
function [png] = cas_pdf()
png.PN_name = 'Using COMMON and specific pre files together';
png.set_of_Ps = {'pS', 'pA', 'pB', 'pC'};
png.set_of_Ts = {'tS', 'tA', 'tB', 'tC'};
png.set_of_As = {'tS','pS',1,... % tS
                 'pS','tA',1, 'tA','pA',1,... %tA
                 'pS','tB',1, 'tB','pB',1,... %tB
                 'pS','tC',1, 'tC','pC',1,... %tC
                };
```

MSF:

```
% Example-08: Using COMMON and specific pre files together
% MSF: cas.m
global global_info
global_info.STOP_AT = 15; % OPTION: stop simulations after 15 TU.

png = pnstruct('cas_pdf');
dyn.ft = {'allothers',1};
pni = initialdynamics(png, dyn);
sim = gpensim(pni);
plotp(sim, {'pS','pA', 'pB', 'pC'});
```

COMMON_PRE:

```
function [fire, trans] = COMMON_PRE(trans)
% COMMON_PRE file codes the enabling conditions

if strcmp(trans.name, 'tS'),
    % if the enabled transition is "tS", just exit COMMON_PRE,
    % as the conditions for firing are coded in its own
    % specific file "tS_pre.m"
    fire = 1;
    return
end
% this is for the three transitions "tA", "tB", and "tC"
n = timesfired(trans.name); % number times transition has fired
fire = lt(n, 3); % fire, if it has already fired less than 3 times
```

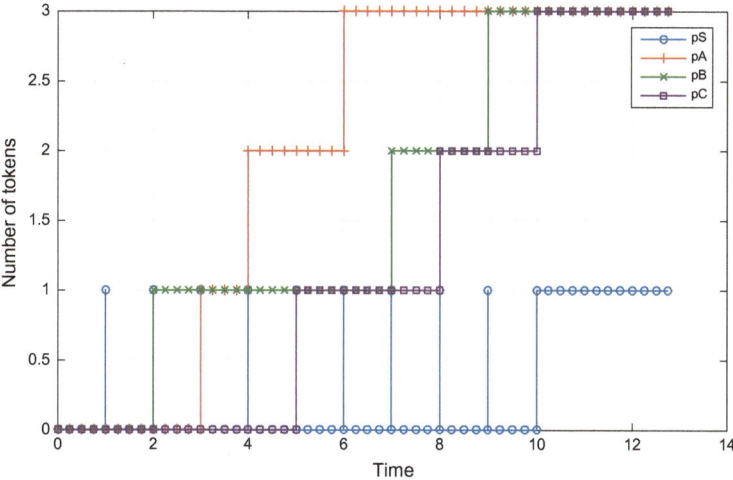

Fig. 3.9 Simulation results

Specific pre-processor for tS: 'tS_pre.m'

```
function [fire, trans] = tS_pre(trans)
% Pre-processor for "tS"
% this is only for "tS"
n = timesfired('tS'); % check how many times "tS" has fired already
fire = lt(n, 10);    % fire, if it has fired less than 10 times
```

Post-processor files: No need as there are no post-firing activities.

Simulation result shows that **tA**, **tB**, and **tC** take three tokens each from **pS** and placing them in pA, pB, and pC, respectively. The tenth token is left in **pS** (Fig. 3.9).

3.8   Combining COMMON and Specific Processors

Let us assume that a Petri net possesses a transition named '**tX**'. Whenever **tX** is enabled, the simulator will execute:

(1) The specific pre-processor 'tX_pre.m', if it exists, and
(2) The COMMON_PRE, if it exists.

Only if both of these pre-processors (if they exist) return **fire = 1**, then **tX** will be allowed to fire.

At the completion of the firing of **tX**, the simulator will execute:

(1) The specific post-processor 'tX_post.m', if it exists, and
(2) The COMMON_POST, if it exists.

3.9   Using Processors as Test Probe

In addition to executing firing conditions (pre-processors) and post-firing actions (post-processors), processors provide a unique functionality: acting as a probe to the simulation engine. Let us explain:

1. The role of Petri net Definition Files (**PDF**): The only use of a PDF is to define a static Petri net graph.
2. The role of Main Simulation File (**MSF**): A PDF will be loaded into memory by MSF right before the simulation start. Thus, an MSF first loads PDF (or PDFs, each PDF representing a module) into the workbench and then starts the simulation. MSF will be blocked during the simulation runs; when the simulation is complete, the control will be passed back to MSF along with the simulation result. Therefore, MSF does not have any control of what going on **during** the simulation.
3. The role of pre- and post-processors: After calling the function 'gpensim' for running the simulation, MSF does not have any control of what is going on during the simulation. However, pre- and post-processors will be automatically called during simulation, *before and after every transition firings*: A pre-processor is to check whether firing conditions of a particular transition are met, and a post-processor is to do some post-firing activities if needed, after the particular transition has fired. Since these are called during the simulations, these can be used to inspect the system run-time, by printing out the values of variables and the parameter values of **PN** and **global_info**.

Bibliographical Remarks
GPenSIM processors (pre- and post-processors) are designed to function as interfaces to deal with the external environment. Davidrajuh (2016a, b) present some work on how GPenSIM processors can be used to interface with some software tools (e.g., Graph Algorithms Toolbox). Davidrajuh (2017) describes how external hardware (e.g., LEGO NXT robot) can be controlled with a Petri net model via the processor files. In this case, the program code for calling the remote procedures and functions for controlling the hardware is coded in the pre- and post-processor files. Chapter 8 presents some more details on using the processors for interfacing with the external environment.

References

Davidrajuh, R. (2016a). Outperforming genetic algorithm with a brute force approach based on activity-oriented petri nets. In *I: International Conference on Soft Computing Models in Industrial and Environmental Applications (SOCO)*, San Sebastian, Spain, October 19–21, 2016.

Davidrajuh, R. (2016b). Detecting existence of cycles in petri nets: An algorithm that computes non-redundant (nonzero) parts of sparse adjacency matrix. In *I: International Conference on Soft Computing Models in Industrial and Environmental Applications (SOCO)*, San Sebastian, Spain, October 19–21, 2016.

Davidrajuh, R. (2017). Petri net based modeling and control of humanoid robots. *International Journal of Simulation, Systems, Science & Technology (IJSSST)*, *17*(32), 40.1–40.9.

Chapter 4
Analysis of Petri nets

This chapter presents two of the functions (tools) that are available in the GPenSIM for the analysis of Petri nets: *coverability tree* and *firing sequences*. In coverability tree analysis, we determine the states that are reachable from a given initial state. The firing sequence is executing a sequence of transitions, to study the state space generated by the firings.

Coverability tree (co-tree) is a very important tool for the analysis of Petri net models. In coverability analysis, we determine the states that are reachable from a given initial state, assuming that all the transitions always fire if enabled. The firing sequence is executing (firing) a sequence of a pre-defined set of transitions, in the strict order of the sequence. This means we force the simulator to allow only the enabled transitions in the pre-defined firing sequence to fire; the other enabled transitions have to wait (will be blocked from firing) until the firing sequence is complete.

4.1 Coverability Tree

This chapter shows how GPenSIM can be used to obtain coverability tree of a Petri net. The methodology for creating a coverability tree of a Petri net is almost the same as for running simulations on a Petri net; the only difference is that in step-3, instead of the function 'gpensim', we use the function 'cotree':

Step-1. Creating Petri net Definition Files (PDFs).

Step-2. No need for pre-processor files, as we assume that transitions always fire, if enabled.

Step-3. Creating Main Simulation File (MSF) with the initial dynamics (only the initial markings; firing times are not relevant); running the MSF using the function '**cotree**' instead of 'gpensim'.

© The Author(s) 2018   43
R. Davidrajuh, *Modeling Discrete-Event Systems with GPenSIM*, SpringerBriefs in
Applied Sciences and Technology, https://doi.org/10.1007/978-3-319-73102-5_4

Function '**cotree**' takes three input arguments (parameters):

1. The first input parameter *pni* is the marked Petri net (with initial markings).
2. The second input parameter *plot_cotree* indicates whether we want to see the graphical plot of the coverability tree.
3. The third input parameter *print_cotree* indicates whether we want to see text printout of the coverability tree.

For example,

```
cotree(pni, 1, 1); % both plot & text printout wanted
```

The above statement will both plot coverability tree graphically as well as print text information on the screen.

Note: For the graphical plot of cotree, we use a modified version of cotree plotting code from the University of Cagliari.

4.1.1 Example-09: Cotree with Finite States

This simple example deals with the Petri net shown in Fig. 4.1. The coverability tree of this Petri net is shown in Fig. 4.2. Let us find the coverability tree using GPenSIM:

PDF:

```
% Example-09: COTREE Example
function [png] = cotree_09_pdf()
png.PN_name = 'COTREE Example-09';
png.set_of_Ps = {'p1', 'p2', 'p3', 'p4'};
png.set_of_Ts = {'t1','t2', 't3'};
png.set_of_As =  {'p1', 't1', 1, 't1', 'p2', 1, 't1', 'p3', 1,...
     'p2','t2',1, 'p3','t2',1, 't2','p2',1, 't2','p4',1,...
     'p3','t3',1, 'p1','t3',1, 'p4', 't3', 1};
```

MSF: The Main Simulation File (after phases 2 and 3) is given below:

```
% Example-09: COTREE Example
pns = pnstruct('cotree_09_pdf');
dyn.m0 = {'p1',2, 'p4',1};
pni = initialdynamics(pns, dyn);
cotree(pni, 1, 1); % cotree: plot graphically and as text
```

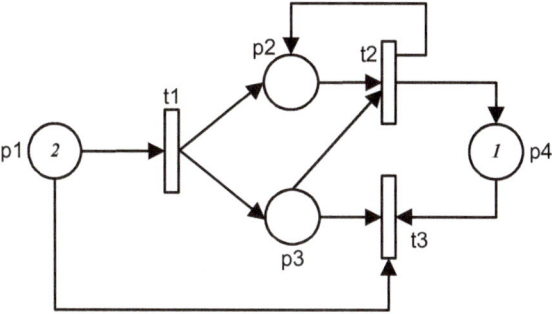

Fig. 4.1 Petri net for coverability analysis

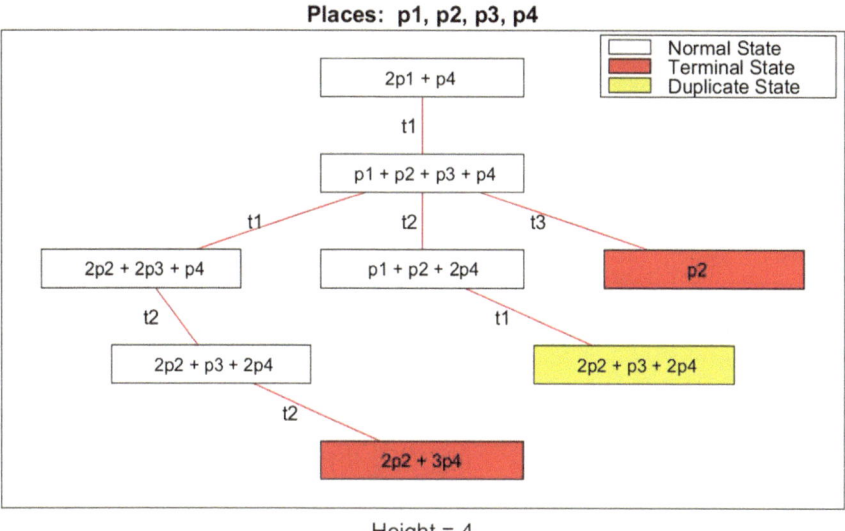

Fig. 4.2 Coverability tree of the Petri net shown in Fig. 4.1

The function **cotree** will plot the cotree graphically (using a modified code from University of Cagliari, for plotting trees) and print text on the screen; both the graphical plot and the text printout are equivalent. However, the printout also prints the boundedness (maximum tokens possible in any place) and the liveness (whether the tree has any terminal nodes) of the Petri net.

Given below is the listing, which is equivalent to the graphical plot shown in Fig. 4.2.

```
======= Coverability Tree =======
State no.: 1   ROOT node
2p1 + p4

State no.: 2     Firing event: t1
State: p1 + p2 + p3 + p4
Node type: ' '   Parent state: 1

State no.: 3     Firing event: t1
State: 2p2 + 2p3 + p4
Node type: ' '   Parent state: 2

State no.: 4     Firing event: t2
State: p1 + p2 + 2p4
Node type: ' '   Parent state: 2

State no.: 5     Firing event: t3
State: p2
Node type: 'T'   Parent state: 2

State no.: 6     Firing event: t2
State: 2p2 + p3 + 2p4
Node type: ' '   Parent state: 3

State no.: 7     Firing event: t1
State: 2p2 + p3 + 2p4
Node type: 'D'   Parent state: 4

State no.: 8     Firing event: t2
State: 2p2 + 3p4
Node type: 'T'   Parent state: 6

Boundedness:
p1 : 2
p2 : 2
p3 : 2
p4 : 3

Liveness:
Terminal States: [5   8]
```

4.1.2   Example-10: Cotree with Infinite States

In this example, we will use GPenSIM to generate coverability tree of the Petri net shown in Fig. 4.3. The coverability tree of this Petri net, which is unbounded, is shown in Fig. 4.4. Let us find the coverability tree using GPenSIM:

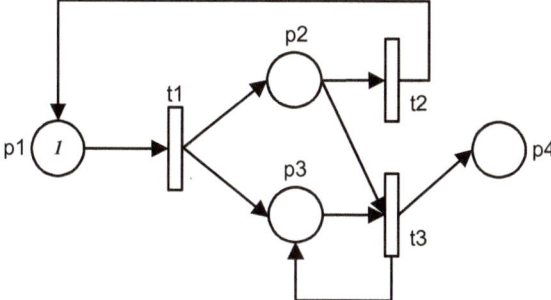

Fig. 4.3 Petri net for coverability tree example-10

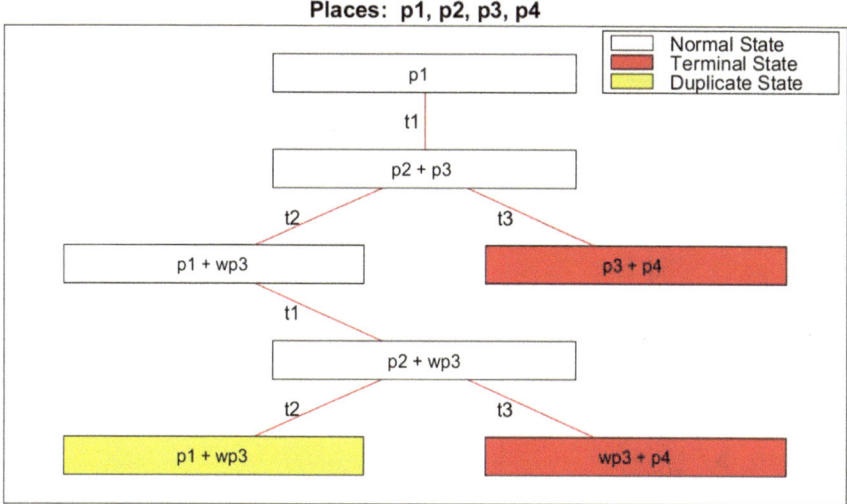

Fig. 4.4 Coverability tree of the Petri net shown in Fig. 4.3

PDF:

```
% PDF Example-10: Cotree example
function [pns] = cotree_10_pdf()
pns.PN_name = 'Petri net for Co-tree example-10';
pns.set_of_Ps = {'p1', 'p2', 'p3', 'p4'};
pns.set_of_Ts = {'t1','t2', 't3'};
pns.set_of_As = {'p1', 't1', 1, 't1', 'p2', 1, 't1', 'p3', 1,...
    'p2','t2',1, 't2','p1',1, 'p2','t3',1 ...
    'p3','t3',1, 't3','p3',1, 't3','p4', 1};
```

MSF:

```
% Example-10: Cotree example
Clear all; clc;
pns = pnstruct('cotree_10_pdf');
dyn.m0 = {'p1',1};
pni = initialdynamics(pns, dyn);
cotree(pns, 1, 1);
```

The print system will print the following on the screen, which is equivalent to
the graphical coverability tree shown in Fig. 4.4.

```
======= Coverability Tree =======
State no.: 1   ROOT node
p1

State no.: 2     Firing event: t1
State: p2 + p3
Node type: ' '   Parent state: 1

State no.: 3     Firing event: t2
State: p1 + Infp3
Node type: ' '   Parent state: 2

State no.: 4     Firing event: t3
State: p3 + p4
Node type: 'T'   Parent state: 2

State no.: 5     Firing event: t1
State: p2 + Infp3
Node type: ' '   Parent state: 3

State no.: 6     Firing event: t2
State: p1 + Infp3
Node type: 'D'   Parent state: 5

State no.: 7     Firing event: t3
State: Infp3 + p4
Node type: 'T'   Parent state: 5

Boundedness:
p1 : 1
p2 : 1
p3 : Inf
p4 : 1

Liveness:
Terminal States: [4   7]
```

The plot system plots the coverability tree shown in Fig. 4.4. In the coverability
tree shown in Fig. 4.4, the 'infinity' is shown as 'w'.

4.2 Firing Sequence

The firing sequence is executing (firing) a sequence of a pre-defined set of transitions,
in the strict order of the sequence. This means we force the simulator to allow only the
enabled transitions in the pre-defined firing sequence to fire; the other enabled transitions have to wait (will be blocked from firing) until the firing sequence is complete.

When we run simulations by calling the function 'gpensim', the maximal set of enabled transitions will be fired at a point of time. Sometimes, we may want controlled firing: We may want to fire a pre-defined series of transitions to see how the system behaves. For example, let us hypothesize that if the transitions **t1**, **t2**, and **t3** are fired one after the other, in that given order, then the system will get into a deadlock situation. If we want to verify this statement, then we may want to fire the sequence {'**t1**','**t2**','**t3**'} in that strict order; the function '**firingseq**' will exactly do that.

The function **firingseq** is a special control function. It interferes with the function **gpensim** so that even if many transitions are enabled at a point of time, only the one that is in the correct position in the pre-defined firing sequence will be considered. If this transition is not enabled, the system will wait until (if possible) it becomes enabled. All other enabled transitions will have to wait until the chosen transition finally fires.

To use the firing sequence, we have to declare some **OPTIONs** in the MSF (OPTIONs are discussed in Chap. 5):

- global_info.FIRING_SEQ: This is the firing sequence of transitions to be fired.
- global_info.FS_REPEAT: How many times the firing sequence has to be repeated; if this option is missing, then the default value of singleton will be set.
- global_info.FS_ALLOW_PARALLEL: Whether to allow a transition to start firing when the previous one(s) is/are still firing (overlapping firings). If this option is missing, then the default value of 'false' will be set.

The function **firingseq** will be called from the pre-processor files (e.g., specific pre or COMMON_PRE). Also, **firingseq** must be used to assign the final value for the output argument 'fire'. If the value for the output argument 'fire' is determined, otherwise the program will malfunction.

Usage in the MSF:

```
global_info.FIRING_SEQ = {'t1', 't2', 't3'}; % the firing sequence
global_info.FS_REPEAT = 3; % repeat firing sequence (t1,t2,t3) 3 times
global_info.FS_ALLOW_PARALLEL = false; % no ovelapping of firings allowed
```

In the pre-processor file:

```
fire = firingseq; % 'fire' must get the value exclusively from 'firingseq'
```

CAUTION! There is only one way to use the function 'firingseq': just assign it to the output variable 'fire' in the pre-processor file. Also, 'fire' must get the value exclusively from 'firingseq'; it is not allowed to combine with other logical conditions.

NOTE: If the transitions given in the firing sequence are not enabled in that order, then indefinite waiting may happen (none of the transition fire).

4.2.1 Example-11: Verifying T-Invariants

In the Petri net shown in Fig. 4.5, the transitions {'**t1**', '**t2**', '**t3**'} are a T-invariant (see Chap. 9 for details on T-invariant and other structural invariants). This means if these transitions are fired one after the other, then we will go back to the original state. Let us verify this statement.

PDF:

```
% Example-11: Firing Sequence
function [png] = firingseq_ex01_pdf()
png.PN_name = 'Example-11: firing Sequence';
png.set_of_Ps = {'p1', 'p2', 'p3', 'p4'};
png.set_of_Ts = {'t1', 't2', 't3'};
png.set_of_As =   {'p4','t1',1, 't1','p1',1, 't1','p2',1, ...   %t1
     'p1','t2',1, 't2','p3',1, ...                    %t2
     'p2','t3',1, 'p3','t3',1, 't3','p4',1}; %t3
```

MSF:

```
% Example-11: Firing Sequence t1, t2, t3
clear all; clc;
global global_info
global_info.STOP_AT = 20;

% OPTIONS for firing sequence
global_info.FIRING_SEQ = {'t1','t2','t3'}; % firing order
global_info.FS_REPEAT = 3;    % repeat the sequence (t1,t2,t3) 3 times
global_info.FS_ALLOW_PARALLEL = false;% no overlapping in firings

pns = pnstruct('firingseq_ex01_pdf');

dyn.m0 = {'p4',5};
dyn.ft = {'t1',1, 't2',2, 't3',3};
pni = initialdynamics(pns, dyn);

prnstate('Initial state: '); % printout the intial state
sim = gpensim(pni);
```

COMMON_PRE:

```
function [fire, trans] = COMMON_PRE(trans)

% assign firingseq directly to 'fire'; otherwise program malfunctions
fire = firingseq();
```

Fig. 4.5 Finding
T-invariants

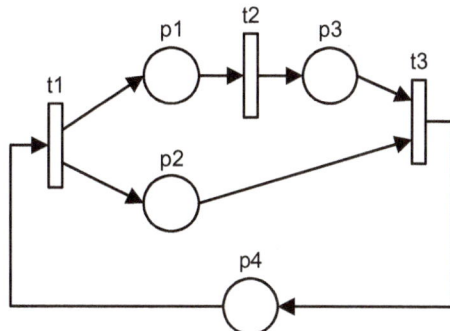

COMMON_POST: Just to print the name of the transition that has fired.

```
function [] = COMMON_POST(trans)
disp(['Fired transition: ', trans.name]); % echo the just fired transition
% at the end of every sequence, printout the current state
if strcmp(trans.name, 't3'),
   prnstate('current state after the firing seq.: ');
end
```

Simulation results:

We see from the results below that at the completion of every firing sequence {**t1**, **t2**, **t3**} the system goes back to the original state, which is five tokens in p4 (**5p4**). Thus, the transitions {'**t1**', '**t2**', '**t3**'} are indeed a T-invariant.

```
Initial state: 5p4

Starting a sequence nr. 1
Fired transition: t1
Fired transition: t2
Fired transition: t3
current state after the firing seq.: 5p4

Starting a sequence nr. 2
Fired transition: t1
Fired transition: t2
Fired transition: t3
current state after the firing seq.: 5p4

Starting a sequence nr. 3
Fired transition: t1
Fired transition: t2

***********************************
Completing the Firing Sequences ....
***********************************

Fired transition: t3
current state after the firing seq.: 5p4
>>
```

4.2.2   Example-12: Alternating Firing with Firing Sequence

In example-05, we made two transitions (**t1** and **t2**) to fire alternatingly using binary semafor. In this section, we will achieve the same results with much less effort! We will use the function **firingseq**, with {'**tX1**', '**tX2**'} as the firing sequence that is to be repeated infinite number of times. In the MSF, we will let the sequence {'**tX1**', '**tX2**'} repeat indefinitely (**global_info.FS_REPEAT = inf**); however, stop the simulations after 10 TU.

MSF:

```
% Example-12: Load Balance with Firing Sequence
clear all; clc;

global global_info
global_info.STOP_AT = 10; % stop the simulations after 10 TU

% data for the strict firing sequence
global_info.FIRING_SEQ = {'tX1', 'tX2'}; % firing sequence
global_info.FS_REPEAT = inf;    % repeat the sequence for ever
global_info.FS_ALLOW_PARALLEL = false; % no parallel firing

pns = pnstruct('firingseq_ex02_pdf');

dyn.m0 = {'pSTART', 10};
dyn.ft = {'tX1',1, 'tX2',2};
pni = initialdynamics(pns, dyn);

sim = gpensim(pni);
plotp(sim, {'p1','p2'});
```

The processor files COMMON_PRE and COMMON_POST become so simple. This is because the whole work is done by the function firingseq! (Fig. 4.6).

COMMON_PRE:

```
function [fire, trans] = COMMON_PRE(trans)
% just call firingseq!
fire = firingseq();
end
```

COMMON_POST (just for echoing name of the fired transition):

```
function [] = COMMON_POST(trans)
% just echo the name of the fired transition
disp(['Fired transition: ', trans.name]);
```

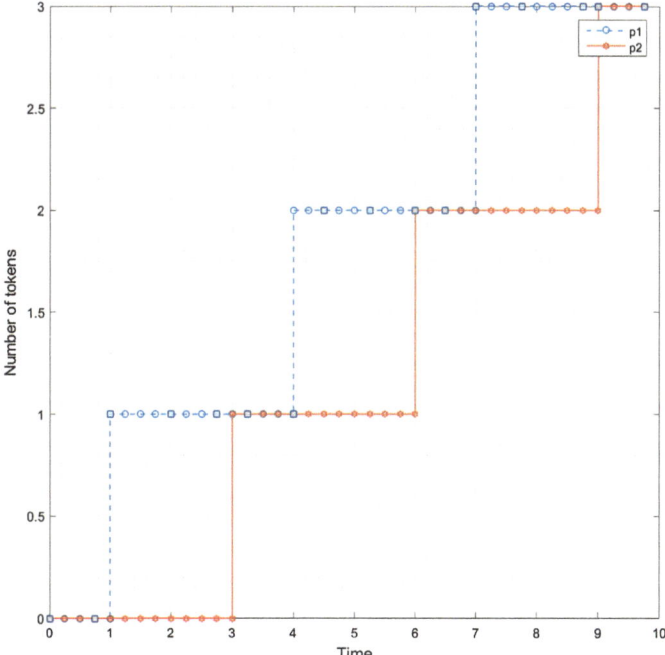

Fig. 4.6 Results of the alternating firing of tX1 and tX2

Simulation results:

```
Starting a sequence nr. 1
Fired transition: tX1
Fired transition: tX2

Starting a sequence nr. 2
Fired transition: tX1
Fired transition: tX2

Starting a sequence nr. 3
Fired transition: tX1
Fired transition: tX2

Starting a sequence nr. 4
>>
```

4.2.3 Example-13: Crosstalk Algorithm

Crosstalk algorithms are an essential part of grid computing, distributed synchro-nizations, and distributed control. In a network of communicating agents, each agent usually has a distinguished initial state; then, it completes a round (usually, sending a message and waiting for acknowledgment), before it starts its next round of operation.

A network of communicating agents is said to run a round policy (or to be round-based) if each message sent in the sender's *i*th round is received in the receiver's *i*th round. Crosstalk arises whenever two agents send each other messages in the same round. Reisig (2013a) presents a crosstalk algorithm, which we shall model and simulate in order to find deadlocks.

Figure 4.7 shows two actor agents who can send and receive messages. The coverability tree (shown in Fig. 4.8) indicates at least one deadlock state [**pLI2 + pLPending + pRI1 + pRPending**], which is reachable by the series of firings, e.g., {**tLAct, tLR2, tRAct, tRL3**}. With the help of **firingseq**, let us verify whether this is true.

> **NOTE:** The coverability tree shown in Fig. 4.8 reveals another problem namely when the number of states of a reachability tree nears fifteen, the reachability tree becomes overcrowded and incomprehensible.

MSF:

```
% Example-13: Crosstalk with 2 Actors, Deadlock prone
global global_info
% data for the strict firing sequence
global_info.FIRING_SEQ = {'tLAct','tLR2','tRAct','tRL3'}; % firing sequence

% the Petri net model is made up of three modules:
%   right module, left module, and the rest (connectors)
pns = pnstruct({'ra_pdf', 'la_pdf', 'connector_pdf'}); % 3 PDFs
dyn.m0 = {'pLQuiet',1, 'pRQuiet',1};
dyn.ft = {'tLAct',1, 'tRAct',1, 'allothers',0.1};
pni = initialdynamics(pns, dyn);
cotree(pni, 1, 1);
sim = gpensim(pni);
prnss(sim);
```

COMMON_PRE:

```
function [fire, trans] = COMMON_PRE(trans)
% assign firingseq directly to 'fire'; otherwise program malfunctions
fire = firingseq();
```

COMMON_POST:

```
function [] = COMMON_POST(trans)
disp(['Fired transition: ', trans.name]); % print the trans just fired
prnstate('Current state: ');   % print current state info
```

Simulation results:

Simulation results clearly indicate that after the firings of the transitions **tLAct**, **tLR2**, **tRAct**, and **tRL3**, the system is deadlocked at time 2.275, as none of the transitions are enabled or firing from this state.

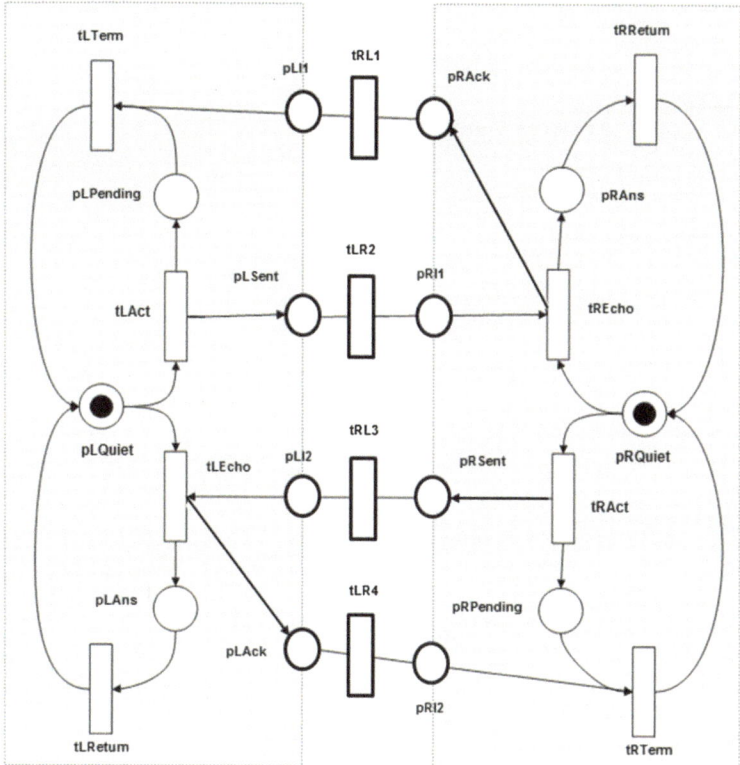

Fig. 4.7 Crosstalk between two agents

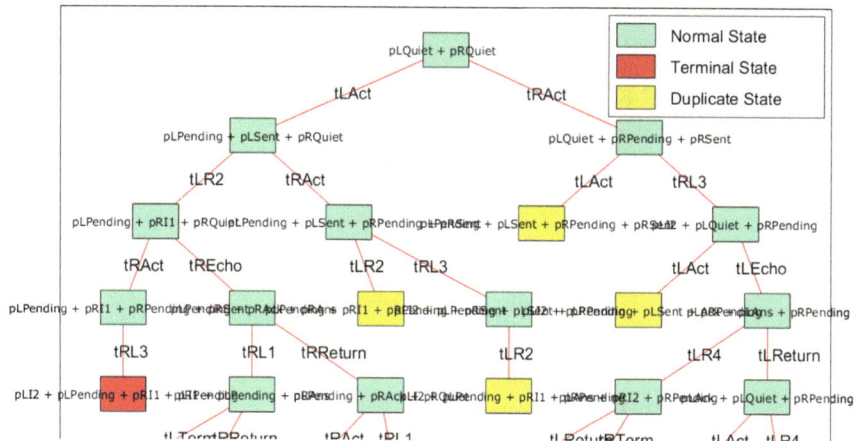

Fig. 4.8 Coverability tree for the crosstalk algorithm

```
Starting a sequence nr. 1
Fired transition: tLAct
Current state:  pLPending + pLSent + pRQuiet

Fired transition: tLR2
Current state:  pLPending + pRI1 + pRQuiet

Fired transition: tRAct
Current state:  pLPending + pRI1 + pRPending + pRSent

Fired transition: tRL3
Current state:  pLI2 + pLPending + pRI1 + pRPending

Simulation of "Crosstalk, Deadlock prone ":
 ======= State Diagram =======
**    Time: 0    **
State:0 (Initial State): pLQuiet + pRQuiet
At start ....
At time: 0,  Enabled transitions are:    tLAct    tRAct
At time: 0,  Firing transitions are:     tLAct

**    Time: 1    **
State: 1
Fired Transition: tLAct
Current State: pLPending + pLSent + pRQuiet
Virtual tokens: (no tokens)

Right after new state-1 ....
At time: 1,  Enabled transitions are:    tLR2    tRAct
At time: 1,  Firing transitions are:     tLR2


**    Time: 1.125    **
State: 2
Fired Transition: tLR2
Current State: pLPending + pRI1 + pRQuiet
Virtual tokens: (no tokens)

Right after new state-2 ....
At time: 1.125,  Enabled transitions are:    tRAct    tREcho
At time: 1.125,  Firing transitions are:     tRAct


**    Time: 2.15    **
State: 3
Fired Transition: tRAct
Current State: pLPending + pRI1 + pRPending + pRSent
Virtual tokens: (no tokens)

Right after new state-3 ....
At time: 2.15,  Enabled transitions are:    tRL3
At time: 2.15,  Firing transitions are:     tRL3


**    Time: 2.275    **
State: 4
Fired Transition: tRL3
Current State: pLI2 + pLPending + pRI1 + pRPending
Virtual tokens: (no tokens)

Right after new state-4 ....
At time: 2.275,  Enabled transitions are:
At time: 2.275,  Firing transitions are:
>>
```

Bibliographical Remarks

In this chapter, reachability tree and coverability tree are treated as the same. However, in some literature, these two are treated as slightly different issues. For example, according to Peterson (1981), the reachability tree represents all possible sequences of transition firings. Thus, the reachability tree can become infinitely large, whereas the coverability tree is the finite-sized abstraction of the reachability tree, with the application of the 'ω' for representing the number of tokens that can grow toward infinity. All the standard textbooks on Petri nets (e.g., Cassandras and LaFortune 2009; Murata 1989; Reisig 2013b) give detailed coverage on coverability tree. Saadallah (2015) presents algorithms for generating coverability trees for different extensions of Petri nets that include inhibitor arcs. Davidrajuh (2013) presents coverability tree that can also reveal the costs of the states in addition to time.

References

Cassandras, C. G., & Lafortune, S. (2009). *Introduction to discrete event systems*. Berlin: Springer Science & Business Media.

Davidrajuh, R. (2013). Extended reachability graph of petri net for cost estimation. In *8th EUROSIM Congress on Modelling and Simulation*, Cardiff, Wales, September 10–12, 2013, pp. 378–382.

Murata, T. (1989). Petri nets: Properties, analysis and applications. *Proceedings of the IEEE, 77* (4), 541–580.

Peterson, J. L. (1981). *Petri net theory and the modeling of systems*. USA: Prentice-Hall.

Reisig, W. (2013a). *Elements of distributed algorithms: modeling and analysis with Petri nets*. Berlin: Springer Science & Business Media.

Reisig, W. (2013b). *Understanding Petri nets: modeling techniques, analysis methods, case studies*. Berlin: Springer.

Saadallah, N. (2015). *Scheduling drilling processes with Petri nets. Ph.D. Dissertation*. Norway: University of Stavanger.

Chapter 5
Optimizing Simulations with GPenSIM

This chapter describes some of the parameters in the GPenSIM that can be fine-tuned for obtaining better simulation results. The parameters that can be adjusted are known as the OPTIONS in GPenSIM. By fine-tuning these OPTIONS, more comprehensible results (e.g., sharper plots, reduced simulation time) can be obtained from the simulations of Petri nets.

'Global info' packet helps passing global variables and parameters between different files (e.g., MSF, PDFs, and processors). In addition, 'Global info' packet also serves another important purpose: setting OPTIONS for simulations. As its name depicts, OPTIONS are not part of the Petri net model; OPTIONS help change default simulation settings. OPTIONS are always stated in capital letters (e.g., 'MAX_LOOP') and added to the 'Global info' packet in the Main Simulation File.

> NOTE: If you use OPTIONS, remember to declare "global_info" as a global variable in the Main Simulation File and also in the other files where you use them.
> E.g.: global global_info % declaration in MSF

5.1 Global Timer

During simulation, GPenSIM uses an internal clock known as the *global timer*; the global timer is discrete and is incremented by a default value that is ¼ of the shortest firing time of any transition. If this default time increment is not satisfactory, then the sampling frequency can be changed (increased) by assigning another value to the timer increment value.

© The Author(s) 2018
R. Davidrajuh, *Modeling Discrete-Event Systems with GPenSIM*, SpringerBriefs in Applied Sciences and Technology, https://doi.org/10.1007/978-3-319-73102-5_5

5.1.1 Example-14: DELTA_TIME Demo

In the figure shown below, let **p1** has three initial tokens. Also, let the firing time of **t1** be 7 s. Though **t1** can fire three times successively, by coding in t1_pre we will force **t1** to fire only at the start of every 30 s (Fig. 5.1).

Since the shortest firing time is 7 s, the sampling rate (timer increment value) is 7/4 = 1.75 s. This sampling rate will be overridden later for better results.

MSF:

```
% Example-14: Timer increment example
global global_info
global_info.STOP_AT = 80;

pns = pnstruct('timer_inc_pdf');
dyn.m0 = {'p1',3};
dyn.ft = {'t1',7};
pni = initialdynamics(pns, dyn);
sim = gpensim(pni);
plotp(sim, {'p1','p2'}, 0, 2);
```

PDF:

```
% Example-14: Timer increment example
function [pns] = timer_inc_pdf()
pns.PN_name = 'Timer increment example';
pns.set_of_Ps = {'p1', 'p2'};
pns.set_of_Ts = {'t1'};
pns.set_of_As = {'p1','t1',1, 't1','p2',1};
```

Pre-processor t1_pre:

```
function [transition] = t1_pre(transition)

% force t1 to fire at the start of every 30 seconds
rest = mod(current_time(), 30);
fire = (rest < 5);    % any number less than 7 would do
```

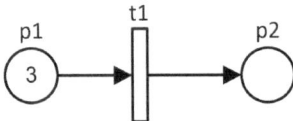

Fig. 5.1 Petri net for the delay example

Simulation results:
If we look closely into the plot for **p1** (Fig. 5.2), the plot changes very slowly when it changes from 2 tokens to 1 token and also from 1 token to 0 tokens. The plot looks like a ramp function (slanted) than an impulse. However, the timer increment (sampling frequency) can be fine-tuned by overriding the default sampling rate, to get a better plot where the plot changes sharply. Let us increase the sampling rate (decrease timer increment value) by overriding the default value for timer increment:

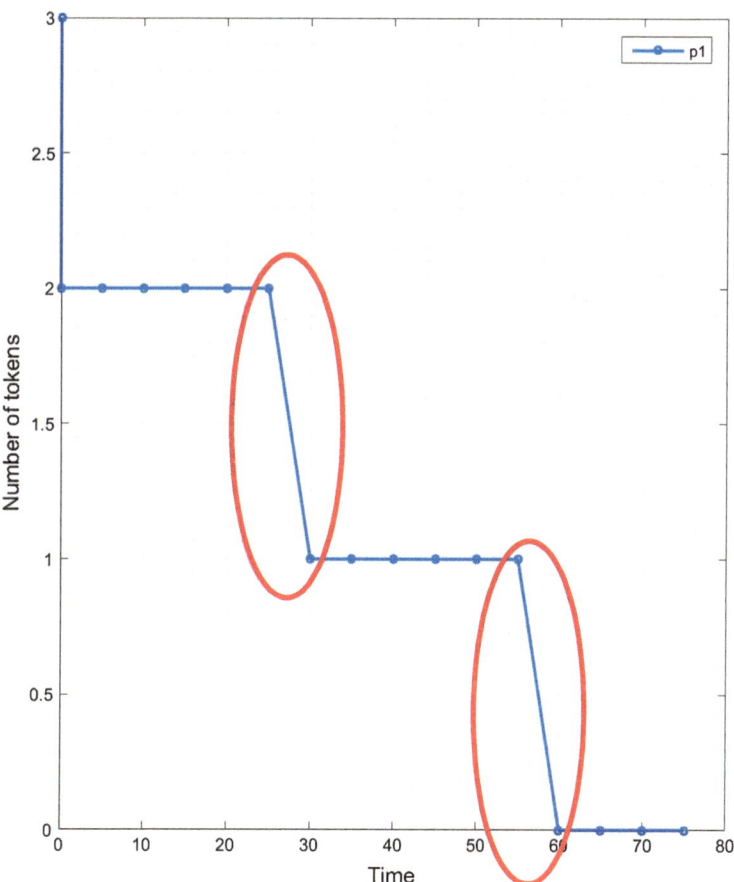

Fig. 5.2 Simulation results

MSF:

```
% Example-14: delay example
% MSF: delay_demo.m
global global_info
global_info.DELTA_TIME = 0.01; % timer increment is decresed to 0.01 sec
global_info.STOP_AT = 80; % stop sim after 80 seconds

pns = pnstruct('delay_demo_pdf');
dyn.m0 = {'p1',3};
dyn.ft = {'t1',7};
pni = initialdynamics(pns, dyn);
sim = gpensim(pni);
plotp(sim, {'p1','p2'}, 0, 2);
```

This means, we have reduced the time increment value from 7/4 s (1.75 s) to 0.01 s, by 175 times; in other words, the sampling frequency is increased 175 times. The simulation result shown below in Fig. 5.3 proves that the sampling is done at a very high rate as the plot for **p1** now looks very sharp like an impulse than a ramp.

Finally, we will decrease the sampling rate and see what happens; let us slow down sampling by assigning a larger value to the timer increment value, say 5 s (again, default increment value is firing time of **t1** divided by 4 = 7/4 = 1.75) (Fig. 5.4):

MSF:

```
% Example-14: timer increment example
global global_info
global_info.DELTA_TIME = 5; % timer increment is 5 sec (low samplings rate)
global_info.STOP_AT = 80;    % stop sim after 80 seconds

pns = pnstruct('timer_inc_pdf');
...
```

NOTE: Increasing the sampling rate will make the simulations run slower.

5.2 'MAX_LOOP'

By default, the simulations are run for **200** cycles or **loops**. Sometimes, running simulation for 200 loops seems unnecessarily lengthy. In this case, 'MAX_LOOP' can be used to trim down the simulation loops to a much lower value (e.g., global_info.MAX_LOOP = 10). However, in some cases, the default 200 loops may seem too little, as the simulation end prematurely. In this case, we may again use the 'MAX_LOOP' OPTION to assign a large value for the number of the simulation loops to be run (e.g., global_info.MAX_LOOP = 20,000).

NOTE: Increase MAX_LOOP for large number of iterations (loops).

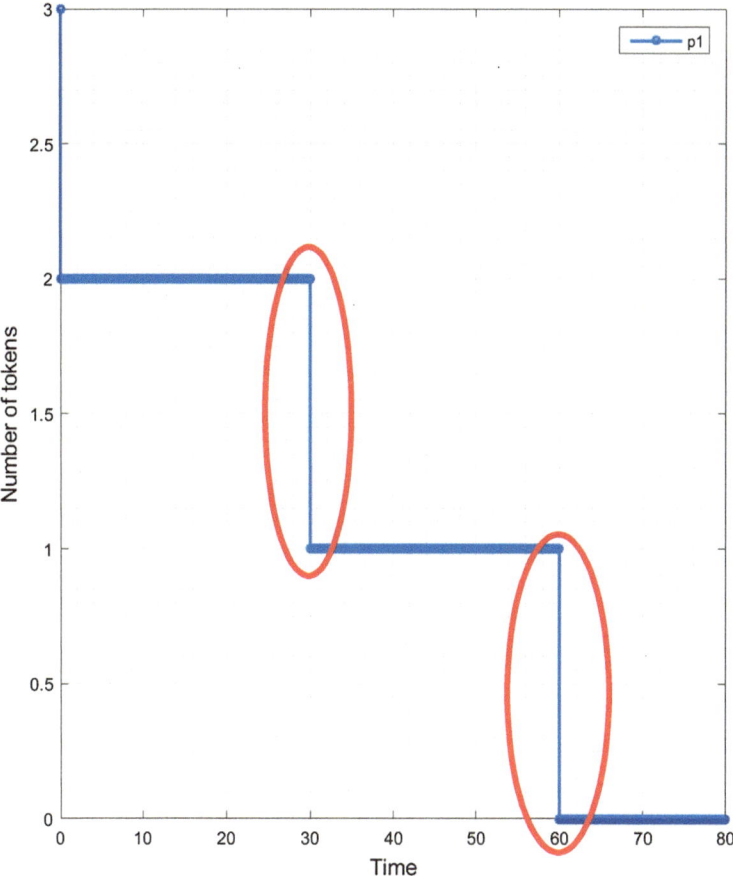

Fig. 5.3 Crisp plots with overridden sampling rate

5.2.1 Example-15: MAX_LOOP

This is the same as the example-01. This time, we will experiment with 'MAX_LOOP' OPTION.

The Petri net shown in Fig. 5.5, with the value of 10 TU as the firing time for **t1**, will only run for a short time. This is because **t1** can fire only two times. Thus, unless specified in the MSF, the default maximum loops of 200 (by default, MAX_LOOP = 200) will be run making the plot (see Fig. 5.6) less detailed.

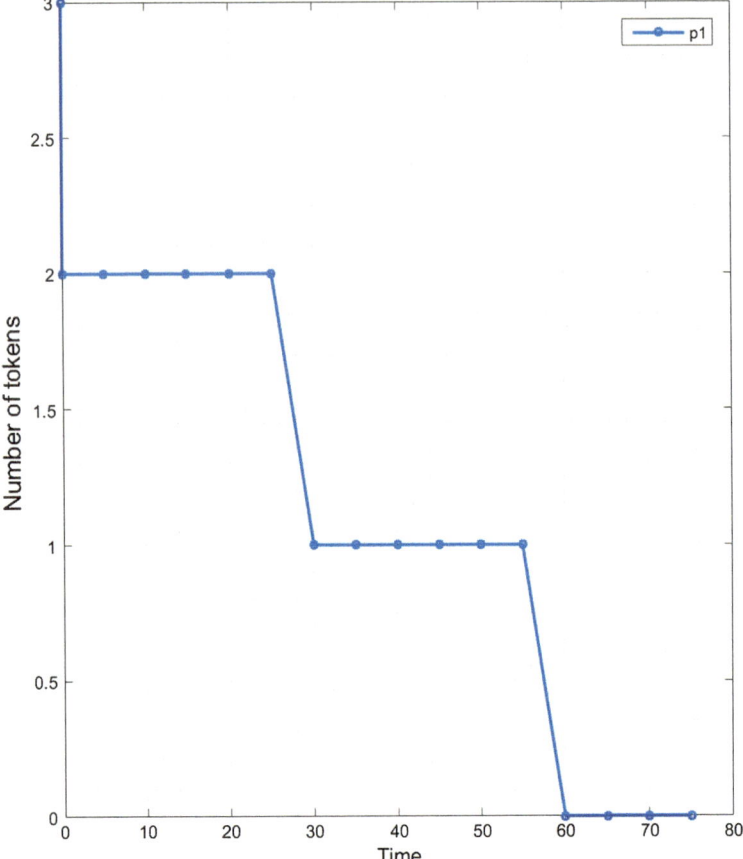

Fig. 5.4 No longer impulse but ramps

Fig. 5.5 Petri net with a
transition that can fire only
twice

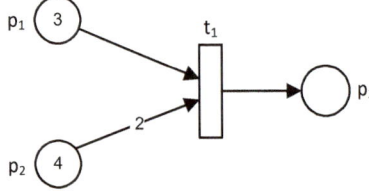

If we stop the simulations after a few simulation loops, then we may have a more
detailed plot. The statement given below limits the simulation loops to 15, by
assigning the value 15 to 'MAX_LOOP':

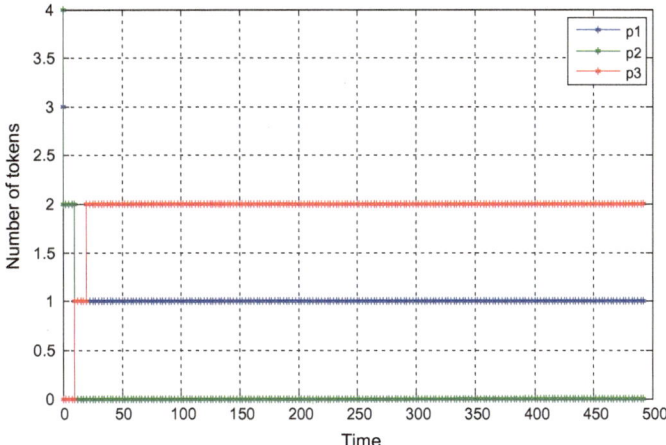

Fig. 5.6 Plot with default MAX_LOOP (=200)

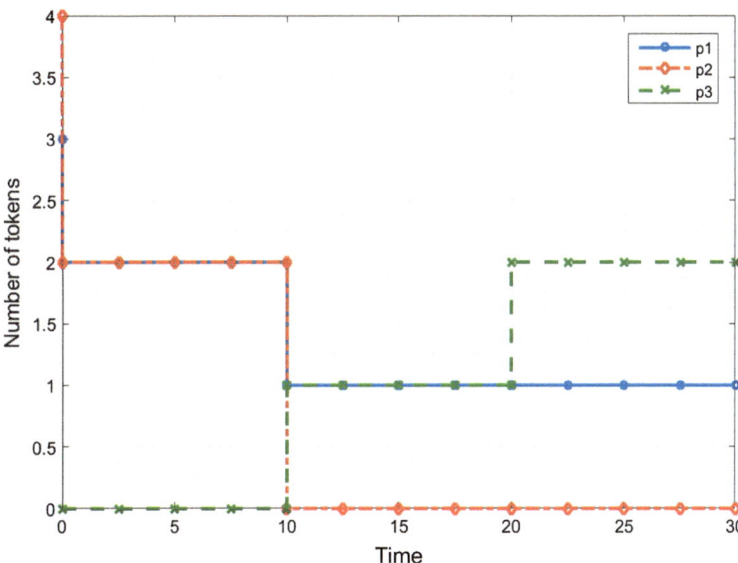

Fig. 5.7 MAX_LOOP is set to 15

```
> global_info.MAX_LOOP = 15; % OPTION to limit simulation loops to 15
```

Simulation results:
When MAX_LOOP is set to 15, a meaningful plot results (Fig. 5.7).

5.2.2   What Are Loops?

To understand loops, we need to understand the theory for general discrete-event simulations (DES). A DES software consists of three main elements (Davidrajuh 2012):

(1) *Global timer*: Global timer (or current time) synchronizes all the activities. The global timer must not be changed by any transitions (events). In GPenSIM, global timer can be accessed in processors, by calling **PN.current_time**, where **PN** is the run-time Petri net structure. Function *current_time()* also returns the current time of the global timer. The global timer is incremented by a discrete value after every cycle, and these cycles are called loops.

(2) *Event scheduler:* Event scheduler is the main activity in a **loop**, performing two actions:

 a. First: checking for any enabled transitions; if there are any enabled transition and if they can fire, then they will be put in *firing_queue* of firing transitions (implemented in the GPenSIM system file **start_firing.m**).

 b. Second: checking for completion of firing of the firing transitions in the *firing_queue*. When the firing of a transition is complete, the transition will be removed from the *firing_queue* (implemented in the GPenSIM system file **complete_firing.m**).

 In GPenSIM, the system file **timed_pensim.m** implements the event scheduler.

(3) *Firing_Queue*: (discussed above).

The loop number states how many cycles of event scheduler have taken place so far.

5.3   'PRINT_LOOP_NUMBER'

When we run simulations of a large Petri net model, we may notice that the **MATLAB hangs**, without giving us any sign of life. It will be better if we can see some outputs during simulations so that we'll be assured that the simulations are going on and that the system is not dead ('hanging'). By setting the OPTION **PRINT_LOOP_NUMBER** to 1 (PRINT_LOOP_NUMBER = 1), we can see the loop numbers when the simulation goes on.

> **NOTE: It is always a good idea to set the 'PRINT_LOOP_NUMBER' to 1 (global_info.PRINT_LOOP_NUMBER = 1) in the MSF.**
> **By setting the PRINT_LOOP_NUMBER = 1, simulation loop number will be displayed during the simulation, and thus we know that simulation is going on and the computer is not 'hanging.'**

5.4 'STARTING_AT'

(Note: For more details, see the section on 'Using Hourly Clock')

So far, we have treated clock as a unitless timer; it will always start at time = 0 during simulation start and will increase afterward. However, there are situations where:

- We need not start at time zero. We may start at time = 600, as things only happen from that time.
- In business modeling applications, it will be much better to use an hourly clock, a clock that uses and shows time in hours, minutes, and seconds. In addition, for business modeling, the clock should perhaps start at 09:00 AM, rather than at '0'.

  OPTION '**STARTING_AT**' is to set the start of the clock to a specific time. For example,

```
global_info.STARTING_AT = [9 0 0]; % start 09:00 AM [09:00:00] HH:MM:SS
```

The time for starting could be given as a three column vector [H M S] or as a single number; if it is given as a single number, then the global timer will keep on running in terms of TUs (e.g., in seconds). Only if the starting time is given as a vector of three elements ([Hour Min Sec]), then the global timer will become an hourly clock.

See the Sect. 5.6 for more details on the hourly clock.

5.4.1 'STOP_AT'

Similar to the OPTION 'STARTING_AT', we can also use 'STOP_AT' to instruct the simulations to stop at a specific point of time. We have already used this option, starting from example-01. 'STOP_AT' OPTION will not work for P/T (untimed) Petri nets as time is undefined (not used) in them.

For example,

```
global_info.STOP_AT = [12 0 0]; % stop sim at 12:00 AM using hour format
```

```
global_info.STOP_AT = 50; % stop sim at 50 TU
```

5.5   'STOP_SIMULATION'

'STOP_SIMULATION' is a special kind of OPTION, in fact the only OPTION that is manipulated in the processor files (pre and post), and not in the MSF; the other OPTIONS are usually set in the MSF, at the top of the file. 'STOP_SIMULATION' is to force the simulations to stop when some specific conditions are met.

5.5.1   Example-16: OPTION 'STOP_SIMULATION' Demo

Figure 5.8 shows a Petri net that will run forever (meaning, up to the default maximum loops). As shown in the example-15, we could reduce the simulation loops by limiting MAX_LOOP (usually for untimed systems) to a lower value; alternatively, we could also assign a value to 'STOP_AT' OPTION (for timed systems only). However, in this example, we will try to stop the simulation run with the 'STOP_SIMULATION' OPTION.

Let us say that we want to stop the simulation after a specific number of **t1** firings: e.g., the simulations must be stopped after three (3) **t1** firings. This can be done simply by checking the **t1** firing count in any pre-processor (either t1_post or t2_post is more suitable, as they are run after t1 firing) and **force the simulation** to stop if the count condition is met. There are two ways to do this:

- Either setting the **STOP_SIMULATION** OPTION in the **t2_pre** after 3 **t1** firing
- Or setting the **STOP_SIMULATION** OPTION in the **t1_post** after 3 **t1** firing.

Fig. 5.8 Live Petri net

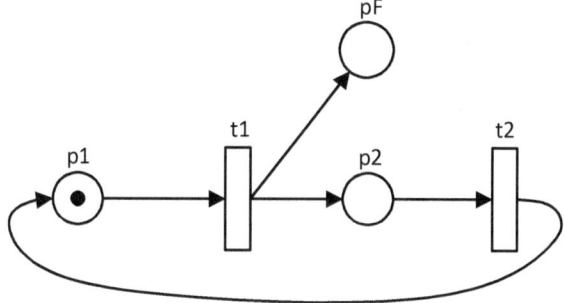

(1) Using the pre-processor t2_pre

```
function [fire, transition] = t2_pre(transition)
global global_info

n1 = timesfired('t1');
if ge(n1, 3),
      % stop simulation, after 3 t1-firing
      global_info.STOP_SIMULATION = 1;
end
fire = 1;
```

Or (2) using the post-processor t1_post

```
function [] = t1_post(transition)
global global_info

n1 = timesfired('t1');
if ge(n1, 3),
    global_info.STOP_SIMULATION = 1;
end
```

MSF:

```
% Example-16: STOP_SIMULATION Demo
clear all; clc;
global global_info

pns = pnstruct('stop_pdf');
dyn.m0 = {'p1',1};
dyn.ft = {'allothers', 10};
pni = initialdynamics(pns, dyn);
sim = gpensim(pni);
plotp(sim, {'pF'}, 0, 2);
```

Results:

The simulation result (Fig. 5.9) shows that the system stops after three **t1**-firings.

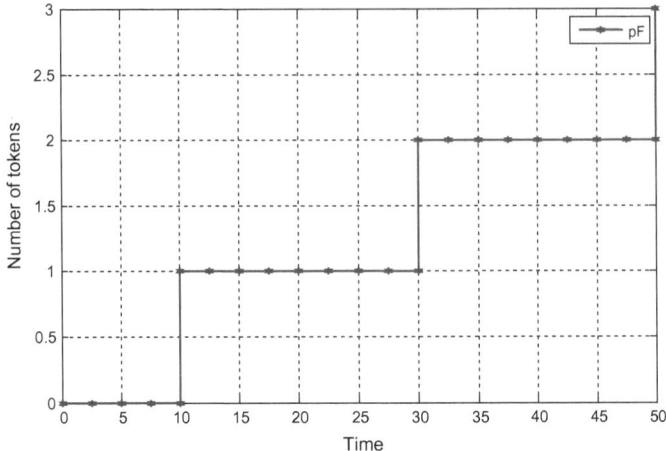

Fig. 5.9 Stopping the simulation

5.6 Using Hourly Clock

So far, we have treated clock as a unitless timer; it will always start at 0 during simulation start and will increase afterward in discrete time intervals. However, in business modeling applications, it will be much better to use an hourly clock, a clock that uses and shows time in hours, minutes, and seconds. The following example explains the issue.

CAUTION! CAUTION!
Time in hourly format must be given as a vector with 3 elements (e.g. 1:00 PM as [13, 0, 0]); we can mix time in 3 column hourly format with a single number; however, single numbers will be taken as seconds.
E.g.:
[0 40 0]   is equivalent to 40 minutes (or 2400 seconds)
'5 + 1*rand(1)'   is equivalent to 5 seconds on average
180   is equivalent to 180 seconds
[9 0 0]   this is 9 hours
34200   this is also 9 hours (32400 seconds = 9x60x60)

5.6.1 Example-17: Hourly Clock

An office opens at 12:00 AM on every business day. Customers arrive at every 15 min. Two clerks will interact with the customers. The clerks take 45 min to service a customer. The office closes at 01:30 PM, and no customer will be allowed into the office. However, those customer(s) already inside the office will be serviced. The Petri net model for this problem is shown in Fig. 5.10.

Fig. 5.10 A simple bank

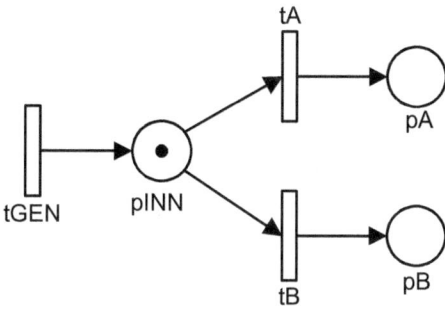

First of all, we want to start the simulation at 12:00 AM. We will use the OPTION '**STARTING_AT**' to set the start time.

```
global_info.STARTING_AT = [12 0 0]; % start 12:00:00 HH:MM:SS
```

In MSF, to assign firing times to **tGEN** representing customer arrival (every 15 min), and to clerk-A and clerk-B (45 min each), we may use either the hourly clock format or times in seconds:

```
dyn.ft = {'tGEN',15*60, 'tA',45*60,'tB', [0 45 0]};
```

Note: Because of the use of hourly clock formats, the functions **prnss** and **plotp** display time information in hourly formats.

In the files shown below, print functions are not used in MSF, as they are coded in the processor files.

MSF:

```
% Example-17: Hourly Clock
global global_info

global_info.START_AT = [12 0 0]; % OPTION: start simulations at 12 AM
global_info.STOP_AT  = [15 15 0]; % OPTION: stop  simulations at 3:15 PM
global_info.DELTA_TIME = 60;  % delta_T is 1 minutes == 60 secs
global_info.BANK_CLOSED = false; % initially, bank is just opened

pns = pnstruct('hourly_clock_pdf');
dyn.m0 = {'pINN',1};   % assume a customer is there at start (12AM)
dyn.ft = {'tGEN',[0 15 0], 'tA', 45*60, 'tB', [0 45 0]};

pni = initialdynamics(pns, dyn);
results = gpensim(pni);
figure(1), plotp(results, {'pINN'}, 0, 2); % number of customers in queue
figure(2), plotp(results, {'pA', 'pB'}, 0, 2); % customers completed
```

PDF:

```
% Example-17: Hourly clock
function [pns] = hourly_clock_pdf()
pns.PN_name = 'Office Hours';
pns.set_of_Ps = {'pINN', 'pA', 'pB'};
pns.set_of_Ts = {'tGEN', 'tA', 'tB'};
pns.set_of_As = {'tGEN','pINN',1, 'pINN','tA',1, 'pINN','tB',1, ...
                 'tA','pA',1, 'tB','pB',1};
```

Pre-processor files:

There is no need for specific pre-processor files for **tA** and **tB**.

However, we need a specific pre-processor for **tGEN** as it should stop generating customers after hours 13:30.

tGEN_pre: tGEN will not fire after 13:30.

```
function [fire, transition] = tGEN_pre(transition)

global global_info

if global_info.BANK_CLOSED, fire = 0; return end

time_stamp = string_HH_MM_SS(current_time()); % convert secs to [hh:mm:ss]
time_13_30_in_seconds = 13.5*60*60;  % Hour 13:30 in seconds

if lt(current_time(), time_13_30_in_seconds),
    disp([time_stamp, ':  ', transition.name, ' Making a customer ']);
    fire = 1;
else
    disp([time_stamp, ':  ',transition.name,' ** BANK IS CLOSED NOW. **']);
    global_info.BANK_CLOSED = true;
    fire = 0;
end
```

COMMON_PRE and COMMON_POST files are just to print statements whenever the clerks take customers and complete business with them.

COMMON_PRE:

```
function [fire, transition] = COMMON_PRE(transition)

time_stamp = string_HH_MM_SS(current_time());% convert secs to [hh:mm:ss]
if ismember(transition.name, {'tA', 'tB'}),
    disp([time_stamp, ':  ', transition.name, ' taking a customer ']);
end
fire = 1;
```

COMMON_POST:

```
function [] = COMMON_POST(transition)

time_stamp = string_HH_MM_SS(current_time());% convert secs to [hh:mm:ss]
if ismember(transition.name, {'tA', 'tB'}),
    disp([time_stamp, ':  ', transition.name, ' finished a customer']);
end
```

Simulation results show that the last customer leaves at 15:00.

```
12:00:00:   tA taking a customer
12:00:00:   tGEN Making a customer
12:15:00:   tB taking a customer
12:15:00:   tGEN Making a customer
12:30:00:   tGEN Making a customer
12:45:00:   tA finished a customer
12:45:00:   tA taking a customer
12:45:00:   tGEN Making a customer
13:00:00:   tB finished a customer
13:00:00:   tB taking a customer
13:00:00:   tGEN Making a customer
13:15:00:   tGEN Making a customer
13:30:00:   tA finished a customer
13:30:00:   tA taking a customer
13:30:00:   tGEN ** BANK IS CLOSED NOW. **
13:45:00:   tB finished a customer
13:45:00:   tB taking a customer
14:15:00:   tA finished a customer
14:15:00:   tA taking a customer
14:30:00:   tB finished a customer
15:00:00:   tA finished a customer
```

5.7   Summary: The OPTIONS

Table 5.1 presents some of the OPTIONS we have seen so far. We will make use of more options later.

5.8   Generators

It is usual to present a generator as shown in Fig. 5.11. However, the model shown in Fig. 5.12 is technically equivalent to Fig. 5.11 and is perfectly capable of continuously generating tokens and supplying the tokens to the rest of the system. This is because the transition **tGEN** in Fig. 5.12 is a cold-start transiton (has no input places) and thus does not need any input tokens to be enabled; hence, **tGEN** in Fig. 5.12 is always enabled.

Token generation need not be either deterministic (firing time of **tGEN** is fixed) or stochastic (firing time of **tGEN** is stochastic, defined as a random function); stochastic firing timing is discussed in Sect. 5.9 'Stochastic Firing Times.' Token generation can be pre-defined, at times pre-defined by the user. The following example shows how this can be achieved.

Table 5.1 OPTIONS

OPTION	Implication

'STARTING_AT'	Declare in the MSF: To set the simulation start time to a specific point of time. For example,
	
	`global_info.STARTING_AT = 60;  % start at 60 TU`
	`global_info.STARTING_AT = [9 0 0];  % start at 9AM`

'STOP_AT'	Declare in the MSF: To stop the simulation at a specific point of time. For example,
	
	`global_info.STOP_AT = 600;  % stop at 600 TU`
	`global_info.STOP_AT = [18 0 0];  % stop at 6PM`

'MAX_LOOP'	Declare in the MSF: To set the simulation loops to a given number. By default, simulations are run 200 loops. For example, `global_info.MAX_LOOP = 20000;  % max 20000 loops`
'PRINT_LOOP_NUMBER'	Declare in the MSF: To print the loop numbers during simulation. For example, `global_info.PRINT_LOOP_NUMBER = 1.` Default value is 0 (loop number printing is switched off)
'DELTA_TIME'	Declare in the MSF: To change the value of global timer advancement (in other words, to change the sampling frequency). E.g.: `global_info.DELTA_TIME = 2;  % timer advancement is 2 TU` The default value is one-fourth of the smallest firing time of any transitions
'STOP_SIMULATION'	Define in pre- or post-processors:: To stop simulation abruptly, when certain conditions are met
'FIRING_SEQ', 'FS_REPEAT', 'FS_ALLOW_PARALLEL'	These options are for executing (firing) a sequence of a pre-defined set of transitions, in the strict sequence order. See the section on 'Firing Sequence'

Fig. 5.11 Generator with
initial enabling token

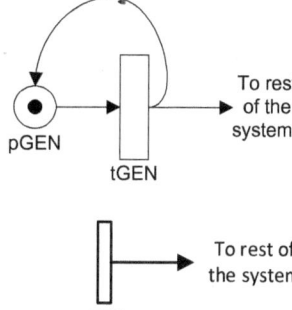

Fig. 5.12 Generator without
initial enabling token

5.8.1   Example-18: Generator for Token Generation

In this example, tokens are generated at specific times. Though these times are read
from a global variable, they can also be read from a structure variable or from a file.

Figure 5.13 shows a simple Petri net. The pre-defined times for firing are fed
into the pre-processor as a global variable.

PDF:

```
% Example-18: Generator example % file: generator_pdf.m: PDF
function [pns] = generator_pdf()

pns.PN_name='Generator Example';
pns.set_of_Ps = {'pOUT'};
pns.set_of_Ts = {'tGEN'};
pns.set_of_As = {'tGEN','pOUT',1};
```

MSF: In the MSF, the pre-defined times for generating tokens are stored in the
global variable '**global_info.tokens_firing_times**'. This variable will be
utilized in the pre-processor.

```
% Example-18: GENERATOR example
global global_info
global_info.STOP_AT = 170;
global_info.tokens_firing_times = [0 20 60 90 100 150];
global_info.DELTA_TIME = 1;

pns = pnstruct('generator_pdf');
% No initial tokens
Dyn.ft = {'tGEN', 1}
pni = initialdynamics(pns, dyn);
sim = gpensim(pni);
plotp(sim, {'pOUT'});
```

Fig. 5.13 Generator (of
tokens)

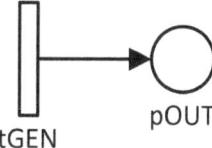

Pre-processor for tGEN:

```
function [fire, transition] = tGEN_pre(transition)
global global_info
% if the variable "tokens_firing_times" is empty, then all
% the firings are done; no more firing is possible
if isempty(global_info.tokens_firing_times),
    fire = 0; return
end

time_to_generate_token = global_info. tokens_firing_times(1);
ctime = current_time();

% if it is time to fire, then remove the time from variable and fire
if ge(ctime, time_to_generate_token),
    if ge(length(global_info.tokens_firing_times),2),
        global_info.tokens_firing_times = ...
                global_info.tokens_firing_times(2:end);
    else
        global_info.tokens_firing_times = [];
    end
    fire = 1;
else  % it is not time to fire
    fire = 0;
end
```

The pre-processor checks the times given in the global variables against the current time and fires if they are equal. After firing, the time is removed from the variable.

Simulation Results:

The plot below shows that firings were done at seconds 0, 20, 60, 90, 100, and 150 (Fig. 5.14).

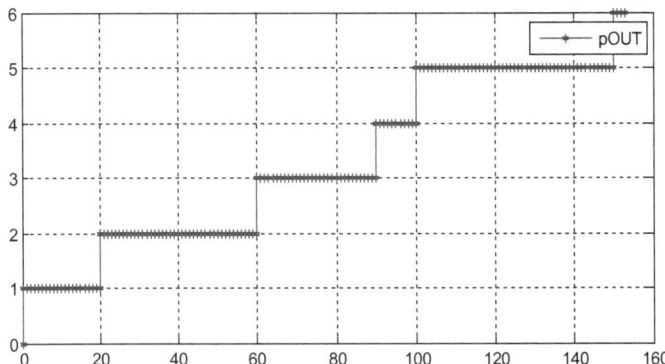

Fig. 5.14 Token generation on pre-defined time

5.9 Stochastic Firing Times and Variable Firing Times

So far, the *firing times* of transitions are assumed to be *deterministic*; thus, the simulations presented so far are deterministic. However, in real-life systems, all the firing times are *stochastic*. GPenSIM provides a limited facility for stochastic firing times.

With MATLAB's Statistics Toolbox, we can use many probability distribution functions. If our computer is *not* installed with Statistics Toolbox, then we have to limit stochastic firing time to basic 'rand' function.

5.9.1 Example-19: Stochastic Firing Times with Advanced Stat. Toolbox

Again, if the computer is installed with MATLAB's Statistics Toolbox, we can use any of the probability distribution functions for stochastic firing times. The following are the most used.

We refer to the CNC production system presented in the example-03; we no longer assume that the firing times are deterministic:

(1) **Robot-1** takes random time binomially distributed with seed 10 and factor 0.9 ms. ('binornd(10,0.9)').
(2) **Robot-2** takes random time normally distributed with mean 1 and standard deviation 0.1 ms. ('normrnd(1,0.1)').
(3) **Robot-3** takes random time uniformly distributed with minimum 8 and maximum 10 ms. ('unifrnd(8,10)').

Thus, the Petri net Definition File is to be changed accordingly:

```
% Example-19: Example with stochastic timing
% the Main Simulation File
pns = pnstruct('stoch_example_pdf');
dynamics.m0 = {'pFrom_CNC', 20}; % initial tokens

% !!! here comes the STOCHASTIC firing times  !!!
dynamics.ft = {'tRobot_1', 'binornd(10,0.9)',...
    'tRobot_2', 'normrnd(1,0.1)', 'tRobot_3', 'unifrnd(8,10)'};

% the code given below is the same as for example-03
pni = initialdynamics(pns, dynamics);
Results = gpensim(pni);
prnss(Results);
plotp(Results, {'pBuffer_1', 'pBuffer_2', 'pBuffer_3'});
```

Note: Due to stochastic timing, up to three different outcomes are possible!!

CAUTION! CAUTION! For defining stochastic firing times with the MATLAB functions like ' `normrnd`', we need the toolbox 'Statistics and Machine Learning' (or 'Advanced Statistics Toolbox' for earlier MATLAB versions).

5.9.2 Example-20: Stochastic Firing Times WO Advanced Stat. Toolbox

If MATLAB's 'Statistics and Machine Learning' toolbox (or 'Advanced Statistical Toolbox' in earlier versions) is not installed on our PC, then we have to use a basic '**rand**' function: For example,

(1) **Robot-1** takes random time normally distributed with mean 10 and standard deviation 2 ms. ('10 + 2*randn(1)').
(2) **Robot-2** takes random time normally distributed with mean 5 and standard deviation 2 ms. ('5 + 2*randn(1)').
(3) **Robot-3** takes random time normally distributed with mean 15 and standard deviation 2 ms. ('15 + 2*randn(1)').

Thus, the Petri net Definition File is to be changed accordingly:

```
% Example-20: Stochastic timing without Adv. Stat. Toolbox
% the Main Simulation File
pns = pnstruct('stoch_example_pdf');
dynamics.m0 = {'pFrom_CNC', 8}; % initial tokens

% !!! here comes the STOCHASTIC TIMING !!!
dynamics.ft = {'tRobot_1', '10 + 2*randn(1)',...
    'tRobot_2', '5 + 2*randn(1)', 'tRobot_3', '15 + 2*randn(1)'};

% the code given below is the same as for example-03
pni = initialdynamics(pns, dynamics);
Results = gpensim(pni);
prnss(Results);
plotp(Results, {'pBuffer_1', 'pBuffer_2', 'pBuffer_3'});
```

Note: Again, due to stochastic timing, up to three different outcomes are possible!!

NOTE: Only the standard (Basic) MATLAB is required to run GPenSIM. GPenSIM does not require MATLAB's Statistics Toolbox or any other toolboxes.

Fig. 5.15 Achieving variable
firing time

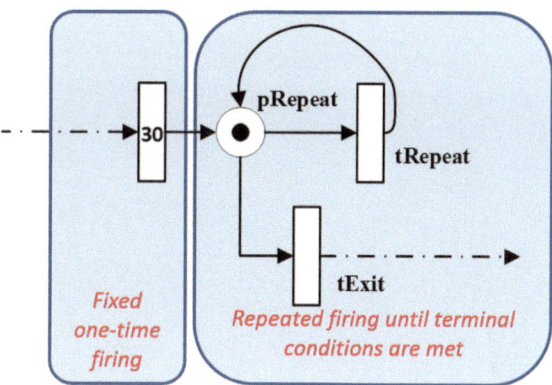

5.10   Achieving Variable Firing Time

In GPenSIM, firing time of a transition is pre-defined, and it can be stochastic or deterministic. A transition cannot have a **variable firing time**. However, variable firing time can be easily simulated if we use a module rather than a single transition.

Let us say that we want a variable firing time between 30 and 40 time units (TUs). As shown in Fig. 5.15, we can achieve this variable firing time with the help of a module which starts with a transition that has the least firing time (30 TU), followed by a repeating block. In this repeating block, transition **tRepeat** (of firing time 1 TU) repeatedly fires until the terminal time has arrived. At this point, the enabling token will be snatched away by **tExit**, making the end of the repeat loop.

Bibliographical Remarks
This chapter presents only the options that are intrinsically available in GPenSIM for improvement of simulations. MATLAB provides some techniques (e.g., profiling) for improving simulations of programs written in MATLAB language. MATLAB Profiling (2016) is to measure how much different parts (functions) of a program take time during the simulation. By profiling, we can see which functions of a program take most of the simulation time and hence try to improve these functions.

References

Davidrajuh, R. (2012). Developing a petri nets based real-time control simulator. *International Journal of Simulation, Systems Science & Technology (IJSST), 12*(3), 28–36.
MATLAB Profiling (2016) https://se.mathworks.com/help/matlab/matlab_prog/profiling-for-improving-performance.html.

Chapter 6
Petri net Extensions and Restrictions

This chapter presents some of the extensions and restrictions to the P/T Petri nets. GPenSIM supports some of these extensions (e.g., colored Petri net, Petri net with inhibitor arc, enabling functions, transitions with priorities) and restrictions (marked graphs, state machines). Due to its flexibility, it is easy to implement the other extensions and restrictions too in GPenSIM.

6.1 Petri net Extensions

P/T Petri net is the original member (or the earliest member) of the Petri net family; though P/T-net is useful as a modeling tool for modeling simple discrete-event systems, it fails to model real-life systems as it lacks modeling power. There are many extensions proposed to increase the modeling power; however, these extensions while increasing the modeling power they also decrease the decision-making (analytical) power (Peterson 1981).

The fundamental extensions to increase modeling power are the *logical extensions* and the *color extension*. Some of the logical extensions are inhibitor arcs, transition priorities, and enabling functions (enabling functions associated with the transitions) (Ciardo 1987); these extensions are discussed in this chapter. Agerwala (1974) shows that Petri net with inhibitor arcs has the same representational power as Turing machines. Ciardo (1987) proves that all three logical extensions—inhibitor arcs, transition priorities, and enabling functions—are equivalent.

Colored Petri net is an extension where the tokens can be loaded with data; tokens loaded with data enable the modeling language to work with systems where synchronization, communication, and resource sharing are important (Saffar et al. 2015). However, Colored Petri net is not discussed in this book, as it is presented in the sequel to this book, Davidrajuh (2018) 'Modeling Discrete-Event Systems with GPenSIM: Advanced Topics.'

© The Author(s) 2018
R. Davidrajuh, *Modeling Discrete-Event Systems with GPenSIM*, SpringerBriefs in Applied Sciences and Technology, https://doi.org/10.1007/978-3-319-73102-5_6

6.2 Petri net Classes

There are several Petri net classes (sometimes referred to as 'subclasses') in use. These classes are based on the restrictions on the input and output places of transitions, or on the input and output transitions of places; for some of the classes, the restrictions are also on the weight of the arcs (Haas 2002). We give below some the classes:

- **Binary Petri net**: A Petri net is called a binary Petri net if all the arcs have unit weight.
- **Petri net State Machine**: A Petri net in which all the transitions have exactly one input place and one output place.
- **Marked Graph (or Event Graph)**: A Petri net in which all the places have exactly one input and one output transitions.
- **Safe Petri nets**: A Petri net in which all the possible markings are one bounded (i.e., number of tokens in any place is at most one).
- **Strongly connected components**: In a strongly connected component, any two nodes are connected in both directions. The issue is to find how many strongly connected components are there in a Petri net.
- **Timed or untimed Petri net**: Whether the transitions in a Petri net posses finite (nonzero) firing time.

It is easy to check the type of a Petri net class with GPenSIM; the function **pnclass** does all the work. This function accepts either the static Petri net graph structure (pns) or the dynamic run-time Petri net structure (pni or PN) as the input argument.

```
pns = pnstruct('stock_pdf');
classtype = pnclass(pns);
```

```
pns = pnstruct('pnclass_demo1_pdf');
dyn.m0 = {'p11',0}; dyn.ft = {'allothers',1};
pni = initialdynamics(pns, dyn);
classtype = pnclass(pni);
```

6.2.1 Example-21: Petri net Class

We are going to check the class type of the Petri net shown in Fig. 6.1. The Petri net is an event graph (more precisely, timed and strongly connected event graph). Let us check the class using the GPenSIM function **pnclass**.

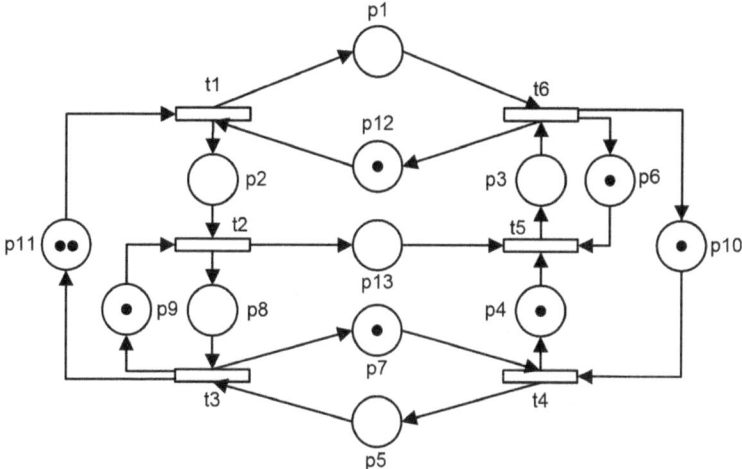

Fig. 6.1 Petri net for checking class type

PDF:

```
% Example-21: PN Classes DEMO - 1
function [pns] = pnclass_demo1_pdf()
pns.PN_name = 'Petri net for Class Demo';
pns.set_of_Ps = {'p1','p2','p3','p4','p5','p6','p7',...
                 'p8','p9','p10','p11','p12', 'p13'};
pns.set_of_Ts = {'t1','t2','t3','t4','t5','t6'};
pns.set_of_As = {...
    'p11','t1',1, 'p12','t1',1,... % input arcs of t1
    't1','p1',1, 't1','p2',1,...   % output arcs of t1
    'p2','t2',1, 'p9','t2',1, ...  % input arcs of t2
    't2','p8',1, 't2','p13',1, ... % output arcs of t2
    'p5','t3',1, 'p8','t3',1, ...  % input arcs of t3
    't3','p7',1, 't3','p9',1, 't3','p11',1,... % output arcs of t3
    'p7','t4',1, 'p10','t4',1,...  % input arcs of t4
    't4','p4',1, 't4','p5',1,...   % output arcs of t4
    'p4','t5',1, 'p6','t5',1, 'p13','t5',1,... % input arcs of t5
    't5','p3',1, ...  % output arcs of t5
    'p1','t6',1, 'p3','t6',1,... % input arcs of t6
    't6','p6',1, 't6','p10',1, 't6','p12',1, ... % output arcs of t5
            };
```

MSF:

```
% Example-21: PN Classes DEMO - 1
pns = pnstruct('pnclass_demo1_pdf');
classtype = pnclass(pns);
```

Results:

```
**** Petri net for Class Demo ****
This is a Binary Petri nets.
This is an Event Graph (Marked Graph).
This is an untimed Petri net.
This is a Strongly Connected Petri net.
>>
```

6.3 Inhibitor Arcs

Petri nets with inhibitor arcs are more powerful than the P/T Petri net. Petri nets with inhibitor arcs have the same expressive power as Turing machines; Turing machines have the ability to model any discrete-event systems. Thus, Petri nets with inhibitor arcs can model any discrete-event systems.

In the Petri net models, an inhibiting arc is drawn with a circle at the end of the arc, instead of pointing arrow.

6.3.1 Firing Rule When Inhibitor Arcs Are Used

A transition is enabled if:

(a) As with P/T Petri nets: The number of tokens in each input place is at least equal to the weight of the input arc from that place.
(b) Special for inhibitor arcs: The number of tokens in each input place with an inhibitor arc is less than the weight of the input inhibitor arc from that place.

Example: In the Petri net shown in Fig. 6.2 (adapted from Ciardo 1987), the transition **t** is enabled if and only if the input place **p3** has greater than or equal to 3 tokens (usual rule) and the inhibiting input places **p1** and **p4** have less than 2 and 4 tokens, respectively.

In the implementation of this Petri net, the only change will be in the Petri net Definition File (PDF). In the PDF, there will be now two types of arcs: the (normal) input arcs and the inhibitor arcs.

PDF:

```
function [pns] = simple_inhib_pdf()
pns.PN_name = 'PDF for: Simple Inhibitor arc Example';
pns.set_of_Ps = {'p1', 'p2', 'p3'};
pns.set_of_Ts = {'t'};
pns.set_of_As = {'p1','t',4};               % normal arcs
pns.set_of_Is = {'p2','t',3, 'p3','t',2}; % Inhibitor arcs
```

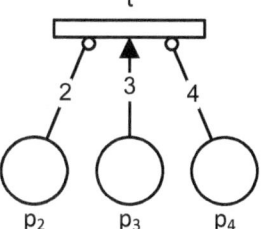

t is enabled *iff* all the following three conditions are satisfied:

$m(p_3) \geq 3$ (normal input)
$m(p_2) < 2$ (inhibiting input)
$m(p_4) < 4$ (inhibiting input)

t is disabled if any of the following three conditions are satisfied:

$m(p_3) < 3$ (normal input)
$m(p_2) \geq 2$ (inhibiting input)
$m(p_4) \geq 4$ (inhibiting input)

Fig. 6.2 Transition *t* with one normal arc (from p_3) and two inhibiting arcs (from p_2 and p_4)

6.3.2 Example-22: Batch Processing

The model shown in Fig. 6.3 describes a simple mechanism for batch processing:

- Producer *tS* produces one item at a time and deposits the products into the buffer *pS*.
- Buffer *pS* has a capacity for holding a maximum of 10 products only; hence, producer *tS* stops when the number of products in the buffer *pS* becomes 10.
- *tE* is the packing machine that picks up 5 products at a time from the buffer *pS*, and packs them as one packet and places it to the output buffer *pE*.
- *tS* takes only 0.2 TU, and *tE* (packaging) takes 5 TUs.

   PDF:

```
% Example-22: A simple Inhibitor Arc example
function [pns] = simple_in_pdf()
pns.PN_name = 'PDF for: Simple Inhibitor arc Example';
pns.set_of_Ps = {'pS', 'pE'};
pns.set_of_Ts = {'tS', 'tE'};
pns.set_of_As = {'tS','pS',1, 'pS','tE',5, 'tE','pE',1}; % input arcs
pns.set_of_Is = {'pS','tS',10};   % !!!! inhibitor arc !!!!
```

   MSF:

```
% Example-22: A simple Inhibitor Arc example
clear all; clc;
global global_info
global_info.STOP_AT = 13;

pns = pnstruct('simple_in_pdf');
dyn.ft = {'tS',0.2, 'tE',5};
pni = initialdynamics(pns, dyn);
sim = gpensim(pni);
plotp(sim, {'pS', 'pE'}, 0, 2);
```

The simulation result is shown Fig. 6.4.

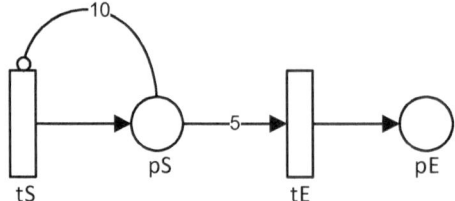

Fig. 6.3 Petri net model of batch processing

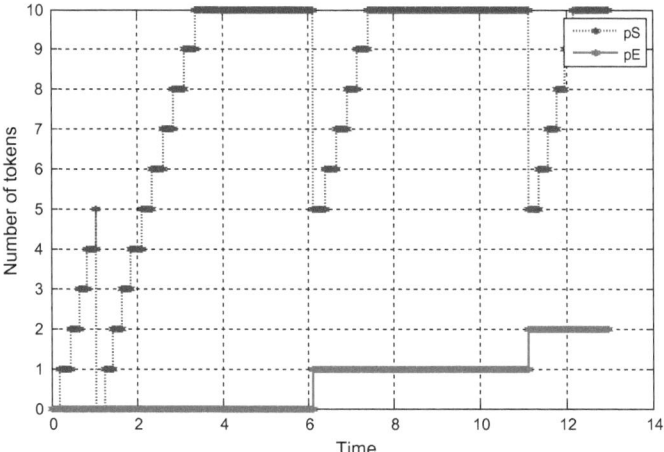

Fig. 6.4 Simulation of batch processing

6.4   Coverability Tree for Petri nets with Inhibitor Arcs

The function for coverability tree 'cotree' discussed in the previous section will not work for Petri nets with inhibitor arcs. This is because the function 'cotree' is based on the 'monotonic' property (monotonic property: If a transition is enabled at the marking x_1, then the transition is enabled for any marking x_2, s.t. $x_2 >= x_1$ (Saadallah 2013). Hence, we have to use another function known as 'cotree**i**' ('i' for inhibitor) for Petri nets with inhibitor arcs.

The function '**cotreei**' takes two input parameters:

1. The first input parameter **pni** is the marked Petri net (with initial markings).
2. The second input parameter is the maximum value for the number of states in the coverability tree.

For example,

```
Cotreei(pni, 15); % print a maximum of 15 states
```

Note: 'cotreei' will not give graphical output. Only the text listing will be printed.

6.4.1   Example-23: Coverability Tree of a Petri net with Inhibitor Arcs

We use the same example discussed in Sect. 6.3.2. However, this time, we only want to see the coverability tree under the influence of the inhibiting arcs.
PDF: same as before

MSF:

```
% Example-23: Coverability tree of a Petri net with
%   Inhibitor Arcs
pns = pnstruct('simple_in_pdf');
dyn.ft = {'tS',0.2, 'tE',5};
pni = initialdynamics(pns, dyn);
cotreei(pni, 15);
```

Results:

```
Warning: initial markings: NOT given ...

 ======= Coverability Tree =======
State no.: 1   ROOT node
(no tokens)

State no.: 2     Firing event: tS
State: pS
Node type: ' '    Parent state: 1

State no.: 3     Firing event: tS
State: 2pS
Node type: ' '    Parent state: 2

State no.: 4     Firing event: tS
State: 3pS
Node type: ' '    Parent state: 3

State no.: 5     Firing event: tS
State: 4pS
Node type: ' '    Parent state: 4

State no.: 6     Firing event: tS
State: 5pS
Node type: ' '    Parent state: 5

State no.: 7     Firing event: tE
State: pE
Node type: ' '    Parent state: 6
```

```
State no.: 8     Firing event: tS
State: 6pS
Node type: ' '   Parent state: 6

State no.: 9     Firing event: tS
State: pE + pS
Node type: ' '   Parent state: 7

State no.: 10    Firing event: tE
State: pE + pS
Node type: 'D'   Parent state: 8

State no.: 11    Firing event: tS
State: 7pS
Node type: ' '   Parent state: 8

State no.: 12    Firing event: tS
State: pE + 2pS
Node type: ' '   Parent state: 9

State no.: 13    Firing event: tE
State: pE + 2pS
Node type: 'D'   Parent state: 11

State no.: 14    Firing event: tS
State: 8pS
Node type: ' '   Parent state: 11

State no.: 15    Firing event: tS
State: pE + 3pS
Node type: ' '   Parent state: 12
>>
```

6.5 Prioritizing Transitions

In discrete systems, we need to increase or decrease the priority of an event or events, in order to prioritize some event(s) and sometimes also to give a fair chance to the competing events. There are some basic facilities in GPenSIM to change priorities of transitions.

(1) *Initial declaration* of priorities, if needed. Initial declaration can only be done in the MSF.
(2) *Increase* priority of a transition (function 'priorinc').
(3) *Decrease* priority of a transition (function 'priordec').
(4) *Get* priority of a transition (function 'get_prior').
(5) *Assign* priority to a transition (function 'priorset').
(6) *Compare* priority of two transitions (function 'priorcomp').

Priority is assigned to transitions only and can be manipulated in processor files. Priority is an integer value (positive, zero, or negative) and higher the priority value the more prioritized the transition becomes. If not given an initial priority value, a transition possesses the default value of 0.

6.5.1   Priorities of Transitions

Assignment of initial priorities can be done in the Main Simulation File; initial priorities can be coded as a part of the initial dynamics.

MSF:

```
dyn.m0 = {'pS', 1}; % initial tokens
dyn.ip = {'t1',3,'t3',5}; % ip stands for initial priority
pni = initialdynamics(pns, dyn);
sim = gpensim(pni);
```

In the above code, we are simply saying that initially, **t1** has a high priority (3), followed by **t3** with the highest priority (5); all other transitions not mention in the declaration will be assigned with the initial priority of zero. When we assign priority, we can assign any integer value, both negative and positive. Higher the value, better the priority is.

Increasing priority of a specific transition can be done using the function '**priorinc**', which will increase the value just by one.

```
priorinc('t1'); % priority of 't1' is increased by one unit
```

Decreasing priority of a specific transition can be done using the function '**priordec**', which will reduce the value by one.

```
priordec('t3'); % priority of 't3' is decreased by one unit
```

We can assign any priority value to a transition using the function '**priorset**':

```
priorset('t1', 10); % priority of 't1' is now 10
```

We can also get current priority of a transition using the function '**get_priority**':

```
pval = get_priority('t1'); % priority value 't1' will be returned
```

Finally, we can compare priorities of two transitions using the function '**priorcomp**':

```
HEL = priorcomp('t3', 't1'); % compare priorities of 't1' and 't3'
```

If **t3** has higher priority than **t1**, then the returned value will be 1; if both have equal priority, then a value of zero will be returned; otherwise, -1 will be returned.

6.5.2 Example-24: Alternating Firing with Priority

In the Petri net model shown in Fig. 6.5, the transitions **t1**, **t2**, and **t3** are supposed
to fire alternatingly. Firing alternatingly can be done by manipulating the priorities
of the transitions:

Let us assume that **t1** is to be fired first, then **t2**, then **t3**, and so on
(**t1** -> **t2** -> **t3** -> **t1** -> **t2** -> ....).

MSF:

```
% Example-24: Alternating firing using priority
global global_info
global_info.STOP_AT = 30;

pns = pnstruct('prio_pdf');
dyn.m0 = {'pS', 1}; % initial tokens
dyn.ft = {'allothers',1}; % firing times of all the trans = 1
dyn.ip = {'t1',1}; % Let t1 fire first, by giving a higher priority
pni = initialdynamics(pns, dyn);

sim = gpensim(pni);
plotp(sim, {'pE1', 'pE2', 'pE3'}, 0, 2);
```

PDF:

```
% Example-24: Alternating firing using priority
function [pns] = prio_pdf()
pns.PN_name = 'Priority Example: Petri net for production facility';
pns.set_of_Ps = {'pS', 'pE1', 'pE2', 'pE3'};
pns.set_of_Ts = {'t1','t2','t3'};
pns.set_of_As = {'pS','t1',1, 'pS','t2',1, 'pS','t3',1,...
     't1','pE1',1, 't1','pS',1, ...
     't2','pE2',1, 't2','pS',1, ...
     't3','pE3',1, 't3','pS',1};
```

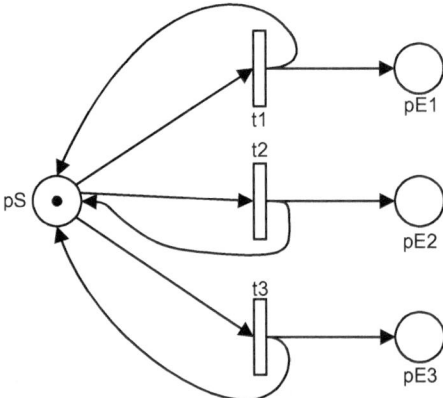

Fig. 6.5 Alternating firing of **t1**, **t2**, and **t3**

COMMON_POST:

We want the following firing sequence: **t1**, **t2**, **t3**, **t1**, …. Hence, we start with the firing of **t1** as we have given a higher priority (value 1) to it in the MSF. When **t1** completes firing, we assign a higher value (priority of **t1** + 1) to **t2** so that **t2** fires next.

```
function [] = COMMON_POST(transition)

% firing sequence: t1 -> t2 -> t3 -> t1 ...
pvalue = get_priority(transition.name); % get priority of the fired trans
switch transition.name
    case 't1'
        priorset('t2', pvalue+1); % give a higher priority to t2
    case 't2'
        priorset('t3', pvalue+1); % give a higher priority to t3
    case 't3'
        priorset('t1', pvalue+1); % give a higher priority to t1
    otherwise
end
```

Simulation Results:

The results show that the three transitions fire alternatingly **t1** -> **t2** -> **t3** -> **t1** … (Fig. 6.6).

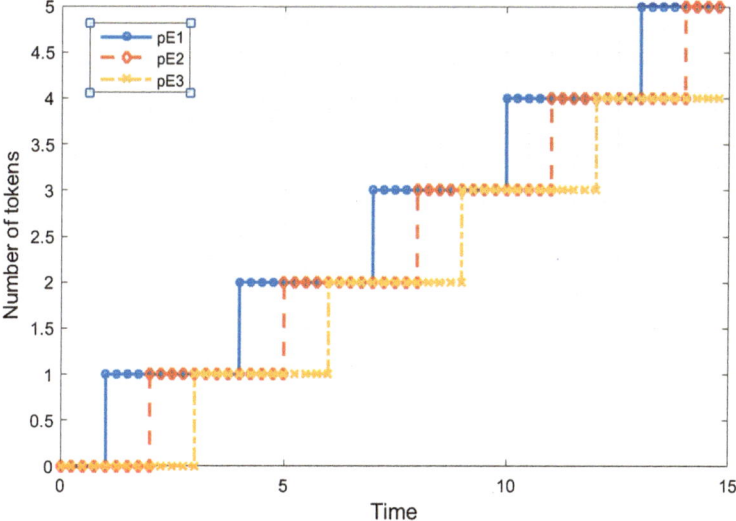

Fig. 6.6 Simulation results for the alternating firing of **t1**, **t2**, and **t3**

6.5.3   Example-25: Priority Decrement Example

This example is the same as the previous example (Example-24). However, this time, we will use priority decrement rather than increment.

MSF:

```
% Example-25: Alternating firing using priority decrement
global global_info
global_info.STOP_AT = 30;

pns = pnstruct('prio2_pdf');

dyn.m0 = {'pS', 1}; % initial tokens
dyn.ft = {'t1',1, 't2',1, 't3',1};
dyn.ip = {'t2',2, 't3',1}; % initial priorities: t2 highest. t1 is lowest.
sim = gpensim(pns, dyn);
plotp(sim, {'pE1', 'pE2', 'pE3'});
```

PDF: same as in the previous example:
COMMON_POST:
Let us say that **t2** has just fired. Then in COMMON_POST, let us assign a lowest value to **t2** (a low value than the other two transitions) so that one of the other two will fire next.

```
function [] = COMMON_POST(transition)

% get the current priority values of the transitions
pvalue1 = get_priority('t1');
pvalue2 = get_priority('t2');
pvalue3 = get_priority('t3');

switch transition.name
% after firing, decrease priority of trans below the other two
    case 't1', pvalue = min(pvalue2, pvalue3) - 1; % t1 is set the lowest p
    case 't2', pvalue = min(pvalue3, pvalue1) - 1; % t2 is set the lowest p
    case 't3', pvalue = min(pvalue1, pvalue2) - 1; % t3 is set the lowest p
    otherwise
end
% set reduced priority to fired trans
priorset(transition.name, pvalue);
```

Simulation Results: Again, same as in the previous example.

Bibliographical Remarks
There are several extensions to Petri nets. GPenSIM supports some of these extensions, such as color (CPN 2017), inhibitor arc (Zaitsev 2013), enabling functions (Gu and Bahri 2002), and transitions with priorities (Guan et al. 1998). At present, GPenSIM does not support the other extensions such as Object-oriented Petri nets (Wang 1996) and Stochastic Petri nets (Bause and Kritzinger 1996). However, due to its flexibility, it is easy to implement Petri net extensions in

GPenSIM, such as Attributed Hybrid Dynamical net (Lopez et al. 2017), Activity-Oriented Petri nets (Skolud et al. 2016, 2017; Krenczyk et al. 2017), and Cohesive Place-Transition Nets with Inhibitor Arcs (Davidrajuh and Saddallah 2016). It is also possible to create modular Petri net models with GPenSIM (Davidrajuh 2017). However, this topic is avoided in this basic book [it is discussed in the sequel to this book, in Davidrajuh (2018)].

In addition to the 'extensions,' we also have the 'restrictions' by which we put some limitation on the Petri net, e.g., on the number of input and output connections, and on the weights of the arcs. The most studied Petri net restrictions are state machines (Murata 1989) and marked graphs (Agerwala 1974). The other well-known restrictions are Free Choice nets (Desel and Esparza 2005), extended Free Choice nets (Desel 1992), and Asymmetric Choice nets (Jiao et al. 2004). All these Petri net restrictions can be studied with GPenSIM.

References

Agerwala, T. (1974). *A complete model for representing the coordination of asynchronous processes*. Baltimore, MD: Hopkins Computer Research Report 32, Johns Hopkins University.

Bause, F., & Kritzinger, P. (1996). *Stochastic Petri nets*. Wiesbaden: Verlag Vieweg. 26.

Ciardo, G. (1987, August). Toward a definition of modeling power for stochastic Petri net models. In *PNPM* (Vol. 87, pp. 54–62).

CPN. (2017). Very brief introduction to CP-nets. Department of Computer Science, University of Aarhus, Denmark.

Davidrajuh, R. (2017). Modular Petri net models of communicating agents. In *International Joint Conference SOCO'17-CISIS'17-ICEUTE'17* León, Spain, September 6–8, 2017, Proceeding (pp. 328–337). Springer, Cham.

Davidrajuh. R. (2018). *Modeling discrete-event systems with GPenSIM: Advanced topics*. Unpublished.

Davidrajuh, R., & Saddallah, N. (2016). Implementation of "Cohesive Place-Transition Nets with Inhibitor Arcs" in GPenSIM. In *IEEE 2016 Asia Multi Conference on Modelling and Simulation*, Kota Kinabalu, Malaysia, December 4–6, 2016.

Desel, J. (1992). A proof of the rank theorem for extended free choice nets. *Application and Theory of Petri nets, 1992*, 134–153.

Desel, J., & Esparza, J. (2005). *Free choice Petri nets*. Cambridge: Cambridge University Press.

Gu, T., & Bahri, P. A. (2002). A survey of Petri net applications in batch processes. *Computers in Industry, 47*(1), 99–111.

Guan, S. U., Yu, H. Y., & Yang, J. S. (1998). A prioritized Petri net model and its application in distributed multimedia systems. *IEEE Transactions on Computers, 47*(4), 477–481.

Haas, P. J. (2002). *Stochastic Petri nets: Modeling Stability*, Simulation, Chapter 8.

Jiao, L., Cheung, T. Y., & Lu, W. (2004). On liveness and boundedness of asymmetric choice nets. *Theoretical Computer Science, 311*(1–3), 165–197.

Krenczyk, D., Davidrajuh, R., & Skolud, B. (2017, September). An activity-oriented petri net simulation approach for optimization of dispatching rules for job shop transient scheduling. In *international joint conference SOCO'17-CISIS'17-ICEUTE'17 León, Spain, September 6–8, 2017, Proceeding* (pp. 299–309). Springer, Cham.

Lopez, F., Barton, K., & Tilbury, D. (2017). *Simulation of discrete manufacturing systems with attributed hybrid dynamical nets.* unpublished in October 2017.

Murata, T. (1989). Petri nets: Properties, analysis and applications. *Proceedings of the IEEE, 77*(4), 541–580.

Peterson, J. L. (1981). *Petri net theory and the modeling of systems.* New Jersey, USA: Prentice-Hall.

Saadallah, N. (2013). Scheduling drilling processes with Petri nets. Ph.D. Dissertation, University of Stavanger, Norway.

Saffar, Y., Jamali, M., & Neshati, M. (2015). Colored Petri nets (CPN). Available: www.cs.ubc.ca/~jamalim/resources/CPN.ppt.

Skolud, B., Krenczyk, D., & Davidrajuh, R. (2016, October). Solving repetitive production planning problems. An approach based on Activity-oriented Petri nets. In *International Joint Conference SOCO'16-CISIS'16-ICEUTE'16* (pp. 397–407). Springer, Cham.

Skolud, B., Krenczyk, D., & Davidrajuh, R. (2017). Multi-assortment production flow synchronization. Multiscale modelling approach. In *MATEC Web of Conferences* (Vol. 112, p. 05003). EDP Sciences.

Wang, L. C. (1996). Object-oriented Petri nets for modelling and analysis of automated manufacturing systems. *Computer Integrated Manufacturing Systems, 9*(2), 111–125.

Zaitsev, D. A. (2013). Toward the minimal universal Petri net. *IEEE Transactions on Systems, Man, and Cybernetics: Systems, 44,* 1–12.

Chapter 7
Performance Evaluation of Discrete-Event Systems

This chapter discusses performance evaluation of discrete-event systems with GPenSIM. The topics such as P/T Petri nets, marked graphs, and strongly connected components are presented. Also, some of the GPenSIM functions that are available for performance evaluation are presented. These functions can be used to find the flow rate, the existence of bottlenecks and deadlocks, etc.

Literature review reveals that the Petri net models of real-life discrete-event systems (e.g., manufacturing systems) are most frequently **marked graphs** (*aka* **event graphs**), a restriction of P/T Petri nets; in addition, they are *strongly connected* event graphs. The literature review also reveals that there are some simple yet powerful methods for performance evaluation and that these methods are only applicable for **strongly connected event graphs**. In this chapter, we look into how we can make use of GPenSIM for performance evaluation.

7.1 Measuring Activation Timing

We are going to find out how much time each transition takes or occupies, out of the total simulation time. From the simulation results, there are two functions that can compute activation time of each transition. Function '**extractt**' creates a simple matrix called the 'duration matrix' in which the first column is the transition (transition index) that fired, the second column is the start time for firing, and the third column is the completion time for firing; thus, the three columns of the duration matrix are:

(1) Column-1: The firing transition (index of the transition)
(2) Column-2: firing start time
(3) Column-3: firing finishing time

Alternatively, we can use the function '**occupancy**' to measure activation times: The function **occupancy** first computes the duration matrix by calling the

© The Author(s) 2018
R. Davidrajuh, *Modeling Discrete-Event Systems with GPenSIM*, SpringerBriefs in Applied Sciences and Technology, https://doi.org/10.1007/978-3-319-73102-5_7

function **extractt**. Then, from the duration matrix, it computes the *occupancy matrix*. Occupancy matrix consists of just two rows:

(1) The first row presents total activation times (in *TUs*) of each transition.
(2) The second row presents activation in the *percentage* of the total time.

The function occupancy also printout the activation times and percentages on screen.

7.1.1 Example-26: Measuring Activation Time

This example is the same as example-14, where the only transition **t1** is forced to wait before each firing. This time, we will compute the idle time of the transition (and the activation time of the transition) with the help of the functions **extractt** and **occupancy** (Fig. 7.1).

The only change from example-14 is that, in this example, there are a few additional lines of code at the end of MSF:

MSF:

```
% Example-26: Delay Example for measuring activation time
global global_info
global_info.STOP_AT = 70;
global_info.DELTA_TIME = 1;

pns = pnstruct('delay_demo_pdf');

dyn.m0 = {'p1',3};
dyn.ft = {'t1',7};
pni = initialdynamics(pns, dyn);
sim = gpensim(pni);

duration_matrix  = extractt(sim, {'t1'});
disp('** Duration_matrix: **');
disp(duration_matrix);

occupancy_matrix = occupancy(sim, {'t1'});
disp('## Occupancy Matrix: ##');
disp(occupancy_matrix);
```

Fig. 7.1 Measuring activation time

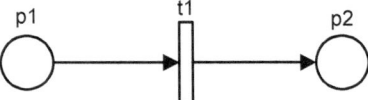

Simulation results:

The duration matrix computed from the simulation result shows that the transition **t1** fired at 0, 30, and 60 time units, and that every firing took 7 time units to complete. Thus, the total time **t1** fired was 21 time units, and the activation percentage was 21/61% = 31.3%. However, the screen dump below indicates that the activation of t1 is 30% as the finishing time for simulation is 70 (=STOP_AT).

```
** Duration Matrix: **
    1    0    7
    1   30   37
    1   60   67

Simulation Completion Time: 70
occupancy t1       :
  total time: 21
  Percentage time: 30%


## Occupancy Matrix: ##
   21
   30
>>
```

7.1.2 Example-27: Measuring Activation Time

This is another example for measuring activation time. Figure 7.2 shows a simple Petri net where two transitions (**t1** and **t2**) fire sequentially, one after the other.

The code below is for the Main Simulation File:

```
% Example-27: Measuring Activation Time
global global_info
global_info.STOP_AT = 507; % stop afer 5 t1/t2 firings

pns = pnstruct('measure_timing_pdf');
dynamicpart.m0 = {'p1', 1};
dynamicpart.ft = {'t1', 1, 't2', 100};
sim = gpensim(pns, dynamicpart);

duartion_martix = extractt(sim, {'t1', 't2'});
disp('Duartion Martix: ');
disp(duartion_martix);


occupancy_martix = occupancy(sim, {'t1', 't2'});
disp('Occupancy Martix: ');
disp(occupancy_martix);
```

Fig. 7.2 Transitions firing
sequentially

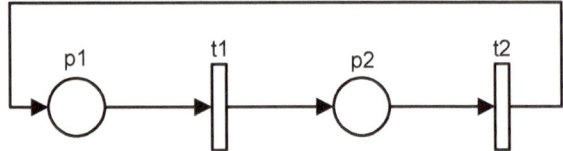

Simulation results:

Simulation result shows that transition **t1** fired during the time intervals 0–1,
101–102, 202–203, ..., etc., and that transition **t2** fired during the time intervals
1–101, 102–202, 203–303, ..., etc. When the simulation was complete at time 506,
t1 has fired six times for a total time of 6 time units, and **t2** has fired five times a
total time of 500 time units.

```
Duartion Martix:
    1       0      1
    1     101    102
    1     202    203
    1     303    304
    1     404    405
    1     505    506
    2       1    101
    2     102    202
    2     203    303
    2     304    404
    2     405    505

Occupancy analysis ....
Simulation Completion Time: 507
occupancy t1       :
  total time: 6
  Percentage time: 1.1834%
occupancy t2       :
  total time: 500
  Percentage time: 98.6193%


Occupancy Martix:
    6.0000   500.0000
    1.1834    98.6193
```

7.2　Minimum Cycle Time in Marked Graphs

Marked graphs (*aka* event graphs) are a class of Petri nets in which *all* the places
have exactly one input and one output transitions; mathematically, $\forall p \in P$: ($\bullet p = 1$)
and ($p \bullet = 1$). For marked graphs, there is an easy way of finding the performance
bottlenecks—the technique is called 'minimum cycle time.' See, for example, Hruz

and Zhou (2007) and Dicesare et al. (1993) for details. The minimum cycle time of a marked graph is given by the equation:

$$\mu = \max_i \frac{D_i}{N_i}$$

where D_i is the total time delay of the *i-th* directed simple cycle (aka elementary circuit), N_i is the total number of tokens in this cycle, and D_i/N_i is the cycle time.

The bottleneck cycle is the *j-th* one where $D_j/N_j = \mu$ holds. To remove the bottleneck, we try to:

- Decrease the D value: This means improving the speed of processes involved in that cycle (reducing the firing times of the respective transitions) and/or
- Increase the N value: Add additional resources (increase the token count) in that cycle.

The technique is explained below with the help of an example.

7.2.1   Example-28: Finding Minimal Cycle Time

Figure 7.3 shows a marked graph as all the places have exactly one input and one output transitions. In addition, it is a strongly connected net. This model consists of eleven **cycles** (*aka* **elementary circuits**). Thus, it is very difficult to find all the eleven cycles manually. Table 7.1 shows the cycles; also shown in the table are the total time delay (TD—summation of the firing times of all the transitions in that cycle), the total number of tokens on each cycle (token sum), and the cycle times.

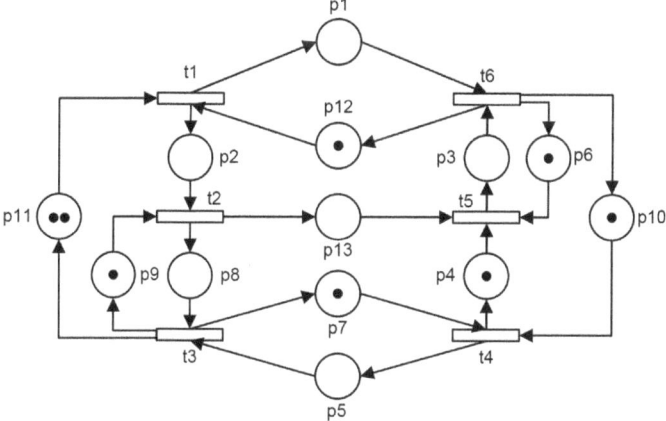

Fig. 7.3 An event graph for finding minimal cycle time

Table 7.1 Finding minimum cycle time

	Cycle	Total TD	Token sum	Cycle time
1.	**p3, t6, p10, t4, p4, t5**	15	2	7.5
2.	**p10, t4, p5, t3, p11, t1, p1, t6**	14	3	4.7
3.	**p13, t5, p3, t6, p10, t4, p5, t3, p11, t1, p2, t2**	21	3	7
4.	**p5, t3, p7, t4**	7	1	7
5.	**t2, p13, t5, p3, t6, p10, t4, p5, t3, p9**	20	2	10
6.	**p3, t6, p6, t5**	11	1	11
7.	**p1, t6, p12, t1**	7	1	7
8.	**t3, p11, t1, p2, t2, p8**	6	2	3
9.	**p3, t6, p12, t1, p2, t2, p13, t5**	14	1	14
10.	**t4, p4, t5, p3, t6, p12, t1, p2, t2, p8, t3, p7**	21	**3**	7
11.	**p8, t3, p9, t2**	5	1	5

From Table 7.1, we see that the bottleneck is the cycle 9 as it takes the most cycle time, which is equal to 14. Thus, if we want to increase the performance of the entire system, we need to concentrate on the elements (transitions) in the cycle 9. We can, either put an extra machine of **t1**, **t2**, **t5**, or **t6** (increase the tokens count N_i) and/or make the machines faster (reduce firing times of **t1**, **t2**, **t5**, or **t6**) in order to reduce the delay D_i.

Let's make use of GPenSIM to find out all these eleven cycles and calculate the other details like total time delay, token sum, and the cycle times.

PDF:

```
% Example-28: Finding Minimum-Cycle-Time in Event Graph
function [png] = event_graph_01_pdf()
png.PN_name = 'Event Graph';
png.set_of_Ps = {'p1','p2','p3','p4','p5','p6','p7',...
                 'p8','p9','p10','p11','p12', 'p13'};
png.set_of_Ts = {'t1','t2','t3','t4','t5','t6'};
png.set_of_As = {...
    'p11','t1',1, 'p12','t1',1,... % input arcs of t1
    't1','p1',1, 't1','p2',1,...   % output arcs of t1
    'p2','t2',1, 'p9','t2',1, ...  % input arcs of t2
    't2','p8',1, 't2','p13',1, ... % output arcs of t2
    'p5','t3',1, 'p8','t3',1, ...  % input arcs of t3
    't3','p7',1, 't3','p9',1, 't3','p11',1,... % output arcs of t3
    'p7','t4',1, 'p10','t4',1,...  % input arcs of t4
    't4','p4',1, 't4','p5',1,...   % output arcs of t4
    'p4','t5',1, 'p6','t5',1, 'p13','t5',1,... % input arcs of t5
    't5','p3',1, ... % output arcs of t5
    'p1','t6',1, 'p3','t6',1,... % input arcs of t6
    't6','p6',1, 't6','p10',1, 't6','p12',1, ... % output arcs of t5
                 };
```

MSF:

```
% Example-28: Finding Minimum-Cycle-Time in Marked Graph
clear all; clc;
pns = pnstruct('event_graph_01_pdf');
dyn.ft = {'t1',1, 't2',2, 't3',3, 't4',4, 't5',5, 't6',6};
dyn.m0 = {'p12',1, 'p6',1, 'p4',1, 'p10',1,...
          'p7',1,'p9',1, 'p11',2};
pni = initialdynamics(pns, dyn);

mincyctime(pni);
```

Simulation results:

```
This is a Strongly Connected Petri net.

Cycle-1:    -> p3 -> t6 -> p10 -> t4 -> p4 -> t5
TotalTD = 15    TokenSum = 2    Cycle Time = 7.5

Cycle-2:    -> p10 -> t4 -> p5 -> t3 -> p11 -> t1 -> p1 -> t6
TotalTD = 14    TokenSum = 3    Cycle Time = 4.6667

Cycle-3:    -> p13 -> t5 -> p3 -> t6 -> p10 -> t4 -> p5 -> t3 -> p11 -> t1
-> p2 -> t2
TotalTD = 21    TokenSum = 3    Cycle Time = 7

Cycle-4:    -> p5 -> t3 -> p7 -> t4
TotalTD = 7    TokenSum = 1    Cycle Time = 7

Cycle-5:    -> t2 -> p13 -> t5 -> p3 -> t6 -> p10 -> t4 -> p5 -> t3 -> p9
TotalTD = 20    TokenSum = 2    Cycle Time = 10

Cycle-6:    -> p3 -> t6 -> p6 -> t5
TotalTD = 11    TokenSum = 1    Cycle Time = 11

Cycle-7:    -> p1 -> t6 -> p12 -> t1
TotalTD = 7    TokenSum = 1    Cycle Time = 7

Cycle-8:    -> t3 -> p11 -> t1 -> p2 -> t2 -> p8
TotalTD = 6    TokenSum = 2    Cycle Time = 3

Cycle-9:    -> p3 -> t6 -> p12 -> t1 -> p2 -> t2 -> p13 -> t5
TotalTD = 14    TokenSum = 1    Cycle Time = 14

Cycle-10:    -> t4 -> p4 -> t5 -> p3 -> t6 -> p12 -> t1 -> p2 -> t2 -> p8 -
> t3 -> p7
TotalTD = 21    TokenSum = 3    Cycle Time = 7

Cycle-11:    -> p8 -> t3 -> p9 -> t2
TotalTD = 5    TokenSum = 1    Cycle Time = 5


Minimum-cycle-time is: 14, in cycle number-9


***  Token Flow Rate:   ***
In steady state, the firing rate of each transition is:
    1/C* = 0.071429
meaning, on average, 0.071429 tokens pass through
any node in the Petri net, per unit period of time.

>>
```

7.3   Expected Performance

We can pass the expected flow rate (tokens per TU) as an optional second parameter
to the function `mincyctime`. In this case, the latter part of the results will change:

In MSF:

```
...
...
pni = initialdynamics(pns, dyn);
mincyctime(pni, 0.1);   % expected flowrate is 0.1 tokens/TU
```

The latter part of the results:

```
...
...
*** We can increase the current flow rate to 0.1 tokens/TU, by improving the
critical circuit alone ...
    In the circuit-9 either:
    1. increase the sum of tokens by 1 tokens, or,
    2. decrease the total delay (firing times) by 4 TU.

>>
```

7.3.1   Example-29: Finding Bottleneck in an Enterprise
Information System

Figure 7.4 shows a Petri net model of an enterprise information system that
involves the following:

- An entry point client request is represented by the transition **tRQ,** and the input
  and output buffers are by the places **pRQi** and **pRQo**. The firing time of **tRQ** is
  2 ms.
- The database server is represented by the transition **tDB**, and the input and
  output buffers are by the places **pDBi** and **pDBo**. The firing time of **tDB** is 6 ms.
- The graphic presentation server is represented by the transition **tGR**, and the
  input and output buffers are by the places **pGRi** and **pGRo**. The firing time of
  tGR is 9 ms.
- The customer resource management (CRM) server is represented by the tran-
  sition **tCR, and** the input and output buffers are by the places **pCRi** and **pCRo**.
  The firing time of **tCR** is 7 ms.
- The place **pDBav** represents the availability of the database server.
- The transitions **tA**, **tB**, and **tC** represent the cost (time) for combining the
  different parts of the reply, with the firing times 3, 2, and 4 msecs, respectively.
- Maximum of two client requests can be processed at a time.

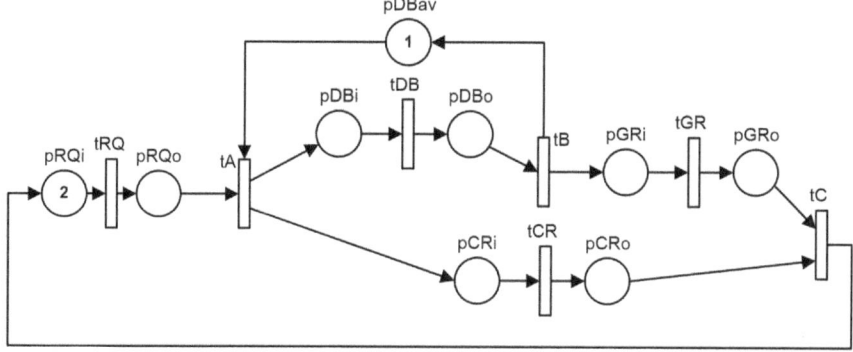

Fig. 7.4 Finding bottleneck in an enterprise information system

PDF:

```
% Example-29: Finding minimum-cycle-time in
%             the Enterprise Information Systems
function [png] = mct_ex02_pdf()
png.PN_name = 'Bottle neck in EIS';
png.set_of_Ps = {'pRQi','pRQo', 'pDBav','pDBi','pDBo',...
                 'pGRi','pGRo', 'pCRi','pCRo'};
png.set_of_Ts = {'tRQ', 'tA','tDB','tB', 'tCR', 'tGR', 'tC'};
png.set_of_As = {...
    'pRQi','tRQ',1, 'tRQ','pRQo',1, ...% tRQ
    'pRQo','tA',1, 'pDBav','tA',1,...  % input arcs of tA
    'tA','pDBi',1, 'tA','pCRi',1,      % output arcs of tA
    'pDBi','tDB',1, 'tDB','pDBo',1, ...% tDB
    'pDBo','tB',1, 'tB','pDBav',1, 'tB','pGRi',1,... %tB
    'pGRi','tGR',1, 'tGR','pGRo',1, ... % tGR
    'pCRi','tCR',1, 'tCR','pCRo',1, ... % tCR
    'pGRo','tC',1, 'pCRo','tC',1, 'tC','pRQi',1, ... % tC
    };
```

MSF:

```
% Example-29: Finding minimum-cycle-time in
%             the Enterprise Information Systems
clear all; clc;
pns = pnstruct('mct_ex02_pdf');
dyn.ft = {'tRQ',2,'tA',3,'tDB',6,'tB',2, 'tGR',9,'tCR',7, 'tC',4};
dyn.m0 = {'pRQi',2, 'pDBav',1};

pni = initialdynamics(pns, dyn);
mincyctime(pni, 0.1); % we want flow rate of 0.1 tokens per mSec
```

Simulation results:

```
This is a Strongly Connected Petri net.

Cycle-1:    -> tA -> pCRi -> tCR -> pCRo -> tC -> pRQi -> tRQ -> pRQo
TotalTD = 16    TokenSum = 2    Cycle Time = 8

Cycle-2:    -> pDBi -> tDB -> pDBo -> tB -> pDBav -> tA
TotalTD = 11    TokenSum = 1    Cycle Time = 11

Cycle-3:    -> tC -> pRQi -> tRQ -> pRQo -> tA -> pDBi -> tDB -> pDBo -> tB
-> pGRi -> tGR -> pGRo
TotalTD = 26    TokenSum = 2    Cycle Time = 13


Minimum-cycle-time is: 13, in cycle number-3


***  Token Flow Rate:   ***
In steady state, the firing rate of each transition is:
    1/C* = 0.076923
meaning, on average, 0.076923 tokens pass through
any node in the Petri net, per unit period of time.


*** We can increase the current flow rate to 0.1 tokens/TU, by improving
the critical circuit alone ...
    In the circuit-3 either:
    1. increase the sum of tokens by 1 tokens, or,
    2. decrease the total delay (firing times) by 6 TU.

>>
```

Bibliographical Remarks

For an extensive coverage of event graphs and finding minimum cycle times, the Chap. 4 'Performance evaluation of manufacturing systems' in DiCesare et al. (1993) is recommended.

References

DiCesare, F., Harhalakis, G., Proth, J. M., Silva, M., & Vernadat, F. B. (1993). *Practice of Petri nets in manufacturing* (p. 8). London: Chapman & Hall.

Hrúz, B., & Zhou, M. (2007). *Modeling and control of discrete-event dynamic systems: With petri nets and other tools*. Berlin: Springer Science & Business Media.

Chapter 8
Interfacing with the External Environment

This chapter discusses interfacing a Petri net model in GPenSIM with the external environment. First, the internal data structures of a Petri net in GPenSIM are described. Also discussed is how the Petri net model developed with GPenSIM can be interconnected with software (e.g., MATLAB Toolboxes) and with hardware (e.g., LEGO Mindstorm robot).

8.1 Internal Data Structures of the Basic Petri net Elements

In this section, we will look into the data structures of some of the basic elements of a Petri net model such as places and transitions. As usual, we will inspect the data structures through an example given below.

> Note: We will see some reference to 'resources' and 'cost' in this section. Resources are an important facility in GPenSIM. However, in this basic book, resources are not dealt with. In addition, cost calculations are also avoided in this book.

8.1.1 Example-30: Data Structures

Figure 8.1 shows a simple Petri net. The model has four places **p1**, **p2**, **p3**, and **p4** and two transitions **tA** and **tB**. We are going to inspect the internal data structures of these elements.

© The Author(s) 2018
R. Davidrajuh, *Modeling Discrete-Event Systems with GPenSIM*, SpringerBriefs in Applied Sciences and Technology, https://doi.org/10.1007/978-3-319-73102-5_8

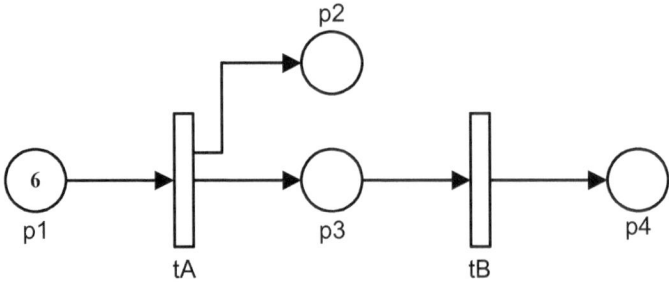

Fig. 8.1 A simple Petri net for inspecting the internal data structures

PDF:

```
% Example-30: data structures of basic PN elements
function [png] = ds_pdf()
png.PN_name = 'ex-30: data structures of basic PN elements';
png.set_of_Ps = {'p1', 'p2', 'p3', 'p4'};
png.set_of_Ts = {'tA', 'tB'};
png.set_of_As = {'p1','tA',1, 'tA','p2',1, 'tA','p3',1,... % tA
                 'p3','tB',1, 'tB','p4',1}; % tB
```

Initial dynamics: **p1** has six initial tokens; firing times of **tA** and **tB** are 1 and 3 TU, respectively.

MSF:

```
% Example-30: data structures of basic PN elements
clear all; clc;
global global_info
global_info.STOP_AT = 24;

pns = pnstruct('ds_pdf');
dyn.m0 = {'p1',6}; % tokens initially
dyn.ft = {'tA',1, 'tB',3}; % firing times

pni = initialdynamics(pns, dyn);
sim = gpensim(pni);
```

Probing the data structures:

When the time becomes 4 TU, **tA** becomes enabled for the fifth time, and the COMMON_PRE will be run for **tA**. At this point, we will interrupt the COMMON_PRE and printout some of the data structures before **tA** starts firing. In order to look into the data structures, we have to make the Petri net run-time structure (**PN**) visible in the COMMON_PRE, by declaring **PN** as a global variable as shown below.

```
function [fire, transition] = COMMON_PRE(transition)

global PN
ctime = current_time;
check_up_time = eq(ctime, 4);

if check_up_time,
    %%%%%%%%%%%%%%%%%%%%
    % Print the data structure of "Petri net run-time", and
    %   of the elements p1, p3, tA, and tB

    disp('The global PN dynamic structure:'); disp(PN);
    disp(' ');

    disp('Structure of the place "p1":');
    index_p1 = is_place('p1');    % get the index of p1
    disp(PN.global_places(index_p1));    % get the data structure of p1

    disp('Structure of the place "p3":');
    index_p3 = is_place('p3');  % get the index of p3
    disp(PN.global_places(index_p3));  % get the data structure of p3

    % look into the data structure of p1 - the token bank details
    disp('Structure of the token bank at "p3":');
    for i = 1:length(PN.global_places(index_p3).token_bank),
        disp(PN.global_places(index_p3).token_bank(i));
    end
    disp(' ');

    disp(' ');
    disp('Structure of the transition "tA":');
    index_tA = is_trans('tA');  % get the index of tA
    disp(PN.global_transitions(index_tA));  % get the data structure of tA

    disp(' ');
    disp('Structure of the transition "tB":');
    index_tB = is_trans('tB');  % get the index of tB
    disp(PN.global_transitions(index_tB)); % get the data structure of tB
end
fire = 1;
```

The simulation results of this example show data structures of some of the basic elements.

8.1.2   The Data Structure of PN—The Global Petri net Run-Time Structure

Given below is the data structure of **PN**—the global Petri net run-time structure. This structure is available everywhere as long as we declare it as a global parameter ('global **PN**') at the beginning of the file (in any specific pre, COMMON_PRE, specific post, and COMMON_POST). PN is a large structure, and it contains all the static as well as dynamic information:

```
The global PN dynamic structure:
                  No_of_modules: 0
                           name: 'example-30: data structur…'
                  global_places: [1x4 struct]
                  No_of_places: 4
            global_transitions: [1x2 struct]
            No_of_transitions: 2
                global_Vplaces: [1x4 struct]
              incidence_matrix: [2x8 double]
          Inhibited_Transitions: [0 0]
              Inhibitors_exist: 0
              inhibitor_matrix: []
            module_membership: [2x2 double]
                  module_names: {}
                        delta_T: 0.2500
                      REAL_TIME: 0
                        HH_MM_SS: 0
                  current_time: 4
                      STOP_TIME: 24
                              X: [2 4 3 1]
                            VX: [0 0 0 0]
          token_serial_numer: 15
        Set_of_Firing_Costs_Fixed: [0 0]
      Set_of_Firing_Costs_Variable: [0 0]
        Firing_Costs_Enabled: 0
            COST_CALCULATIONS: 0
          Set_of_Firing_Times: [1 3]
                  priority_list: [0 0]
              system_resources: []
        No_of_system_resources: 0
            Resource_usage_LOG: []
                    PRE_exist: [0 0]
                    POST_exist: [0 0]
                MOD_PRE_exist: [1x0 double]
              MOD_POST_exist: [1x0 double]
                  COMMON_PRE: 1
                  COMMON_POST: 0
            Firing_Transitions: [0 0]
          Enabled_Transitions: [1 1]
                        State: 5
              attempting_trans: 1
              completing_trans: 2
```

8.1.3   The Data Structure of Place

At time equals to 4 TU, the places **p1** and **p3** have two and three tokens, respectively. Given below is the overall data structure for a place, with values filled in for **p1** and **p3**, at the time 4 TU:

```
Structure of the place "p1":
            name: 'p1'
          tokens: 2
     token_bank: [1x2 struct]

Structure of the place "p3":
            name: 'p3'
          tokens: 3
     token_bank: [1x3 struct]
```

The structure for **p3** has three tokens, and the token bank contains the details about the three tokens, as shown below. The values shown below states that the first token was put into **p3** at time equals to 2 TU ('creation_time = 2'), the second at time equals to 3 TU, and so on. Since we did not assign any costs for transition firing, all the tokens were made with the same cost of 0 cost units, and none of them have colors.

```
Structure of the token bank at "p3":
            tokID: 10
    creation_time: 2
            color: {0x1 cell}
             cost: 0

            tokID: 12
    creation_time: 3
            color: {0x1 cell}
             cost: 0

            tokID: 14
    creation_time: 4
            color: {0x1 cell}
             cost: 0
```

8.1.4   The Data Structure of Transition

The data structure of the transitions **tA** and **tB** at time equal to 4 TU is shown below.

```
Structure of the transition "tA":
                name:  'tA'
         firing_time:  1
         firing_cost:  0
         times_fired:  4
      absorbed_tokens:  [0 0 0 0]


Structure of the transition "tB":
                name:  'tB'
         firing_time:  3
         firing_cost:  0
         times_fired:  1
      absorbed_tokens:  [0 0 0 0]
```

8.2 PNML-2-GPenSIM Converter

The Petri net Markup Language (PNML) is an XML-based transfer format for Petri nets so that the Petri net model can be freely transferred from one tool to another (PNML 2017). PNML is currently standardized in ISO/IEC-15909. This section shows how the PNML capability enhances the usefulness of GPenSIM. This section proposes an approach that eliminates the need for integrated GUI (graphic editor) for GPenSIM. The approach consists of the following three steps:

1. Basic Animations and Simulation: The initial Petri net model can be defined using a graphical Petri net editor like PIPE2 (PIPE2 2017). Using the same tool, basic token game animations can also be done to verify whether the crude model behaves as it should be. Once the modeler is satisfied with the crude model, he can save the model as a PNML file.

2. Importing the initial Petri net model into GPenSIM: PNML file is imported into GPenSIM environment, using the PNML-2-GPenSIM converter. The PNML-2-GPenSIM converter extracts the static Petri net structure and saves it as a PDF, and the initial dynamics (initial markings and the firing times of the transitions) are saved in the MSF.

3. Advanced modeling and simulation: Once the modeler is satisfied with the basics simulations in GPenSIM, enabling functions, transition priorities, resources, and any advanced facilities available in GPenSIM can be coded the processor files (pre and post) to make the advanced model.

The GPenSIM function **pnml2gpensim** is the PNML-2-GPenSIM converter that reads a PNML document describing a Petri net model, extracts the Petri net structure, and then creates PDF and MSF files representing the model. During this process, the graphical details coded in the PNML file are discarded.

Usage:

```
pnml2gpensim(PNMLFile);
```

MATLAB offers a set of functions for reading and interpreting XML files, starting with the function 'xmlread' that reads an XML document and returns Document Object Model (DOM) node. From the DOM node, the elements of the node (such as 'place,' 'transition,' and 'arc') can be visited recursively, extracting the names of the elements, the initial marking (in case of place element), and the source and the target (in case of an arc element). The following steps are involved in the GPenSIM function **pnml2gpensim**:

1. Convert XML file to MATLAB structure and get the root of the DOM tree.
2. From the root tree, recursively visit the child nodes to get the PNML structure.
3. From the PNML structure, get the 'net' child and start extracting Petri net structure (places, transitions, and arcs).
4. Write the Petri net structure into the GPenSIM files MSF and PDF.

8.2.1   Example-31: Generating GPenSIM Files from a PNML File

Let us assume that we have generated a PNML file, named 'fms.xml' with the help of a Petri net tool [e.g., PIPE2 (PIPE2 2017)]. The file 'samplePNML1.xml' is shown in Fig. 8.2. Just passing this file to the function **pnml2gpensim** will produce all the necessary GPenSIM files MSF and PDF; in addition, a template for COMMON_PRE and COMMON_POST will be also generated.

Generating GPenSIM files from PNML File:

```
% Example-31: Convert PNML file into GPenSIM files
clear all; clc;

PNMLfile = 'samplePNML1.xml';
pnml2gpensim(PNMLfile);
disp(' ');
disp(' ************** ');
disp('GPenSIM files are generated for the PNML file: ');
disp(['           ', PNMLfile]);
```

```xml
  <?xml version="1.0" encoding="ISO-8859-1"?>
- <pnml>
    - <net type="P/T net" id="Simple-Petri-Net">
        + <place id="p1">
        - <place id="p2">
            - <graphics>
                  <position y="300.0" x="100.0"/>
              </graphics>
            - <name>
                  <value>p2</value>
                  <graphics/>
              </name>
            - <initialMarking>
                  <value>1</value>
                - <graphics>
                      <offset y="0.0" x="0.0"/>
                  </graphics>
              </initialMarking>
          </place>
        + <place id="p3">
        - <transition id="t1">
            - <graphics>
                  <position y="200.0" x="200.0"/>
              </graphics>
            - <name>
                  <value>t1</value>
                  <graphics/>
              </name>
            - <orientation>
                  <value>0</value>
              </orientation>
            - <rate>
                  <value>1.0</value>
              </rate>
            - <timed>
                  <value>false</value>
              </timed>
          </transition>
        + <arc id="p1 to t1" target="t1" source="p1">
        - <arc id="p2 to t1" target="t1" source="p2">
              <graphics/>
            - <inscription>
                  <value>1</value>
                  <graphics/>
              </inscription>
              <arcpath id="000" y="295" x="105" curvePoint="false"/>
              <arcpath id="001" y="210" x="195" curvePoint="false"/>
          </arc>
        + <arc id="t1 to p3" target="p3" source="t1">
      </net>
  </pnml>
```

Fig. 8.2 Sample PNML file

The generated MSF will be named 'msf.m' and the PDF will be 'pdf_pdf.m'.
Automatically generated 'msf.m':

```
% GPenSIM Main Simulation File
% this MSF is generated from PNML file "samplePNML1.xml"
% MSF: 'msf.m'
clear all; clc;
global global_info % global user data attached to global_info
global_info.PRINT_LOOP_NUMBER=1; % show loop numbers during simulation

pns = pnstruct('pdf_pdf');
dyn.m0 = {'P0',5, 'P2',3, 'P3',2};

pni = initialdynamics(pns, dyn);
sim = gpensim(pni);

prnss(sim);
```

Automatically generated 'pdf_pdf.m':

```
% GPenSIM PDF file generated from PNML file "samplePNML1.xml"
% PDF: 'pdf_pdf.m'

function [png] = pdf_pdf()

png.PN_name = 'PDFxxx';
png.set_of_Ps = {'P0','P1','P2',...
          'P3','P4'};
png.set_of_Ts = {'T0','T1','T2',...
          'T3'};
png.set_of_As = {'P0','T0',1, 'P1','T1',1, ...
       'P2','T0',1, 'P2','T2',3, 'P3','T2',1, ...
       'P4','T3',1, 'T0','P1',1, 'T1','P0',1, ...
       'T1','P2',1, 'T2','P4',1, 'T3','P2',3, ...
       'T3','P3',1};
```

Automatically generated 'COMMON_PRE.m':

```
% COMMON_PRE file generated from PNML file "samplePNML1.xml"
% 'COMMON_PRE.m'

function [fire, transition] = COMMON_PRE(transition)
%function [fire,transition] = COMMON_PRE(transition)

if (strcmpi(transition.name, 'T0')),
elseif (strcmpi(transition.name, 'T1')),
elseif (strcmpi(transition.name, 'T2')),
elseif (strcmpi(transition.name, 'T3')),
else
    % error (['Error in the transition name: ', transition.name]);
end

% fire = 1; % let it fire
```

Automatically generated 'COMMON_PRE.m':

```
% COMMON_POST file generated from PNML file "samplePNML1.xml"
% 'COMMON_POST.m'

function [] = COMMON_POST(transition)
%function [] = COMMON_POST(transition)

if (strcmpi(transition.name, 'T0')),

elseif (strcmpi(transition.name, 'T1')),

elseif (strcmpi(transition.name, 'T2')),

elseif (strcmpi(transition.name, 'T3')),

else
    % error (['Error in the transition name: ', transition.name]);
end
```

8.3   Avoiding PDF Files

Implementing a Petri net model in GPenSIM starts with the PDF file. However, coding a PDF can be boring and sometimes error-prone. Writing a PDF also can be laborious too.

We may want to skip writing a PDF, if we don't care about the names of the places and transitions. We don't need to write a PDF if the PDF file is too large and has cycles on it. GPenSIM provides a function called **'createPDF'** that can automatically create a PDF file for us. If we provide just two matrices, *Ai* (the input incidence matrix) and *Ao* (the output incidence matrix), then **createPDF** will automatically create a PDF file for us.

Usage:

```
createPDF(Ai, Ao, 'PDF_Filename', 'Petri_net_name');
```

The first two input parameters *Ai* and *Ao* are compulsory:

- *Ai* is the input incidence matrix and *Ao* is the output incidence matrix; both of them are of dimension $(m \times n)$, where m is the number of transitions and n is the number of places.

The second and third input parameters are optional:

- PDF_Filename: We can assign a name for the PDF to be created by the function, for example, 'simple_pdf.m'. If name for the PDF is not given, then the default name 'def*nnn*_pdf.m' (where *nnn* is a random number) will be assigned for the PDF.

- Petri_net_name: We can also give a name for the Petri net model, for example, 'This is a simple Petri net'. If the name for the Petri net model is not given, then the default name 'defxxx' will be given.

In the PDF created, all the places and the transitions will be named in numerically increasing order, for example, **'p1'**, **'p2'**, ... , and **'t1'**, **'t2'**, ...

8.3.1 Example-32: A Simple Model Without PDF

Figure 8.3 shows a simple Petri net that consists of six places and four transitions. The input incidence matrix *Ai* and the output incidence matrix *Ao* of this Petri net are shown in Fig. 8.4.

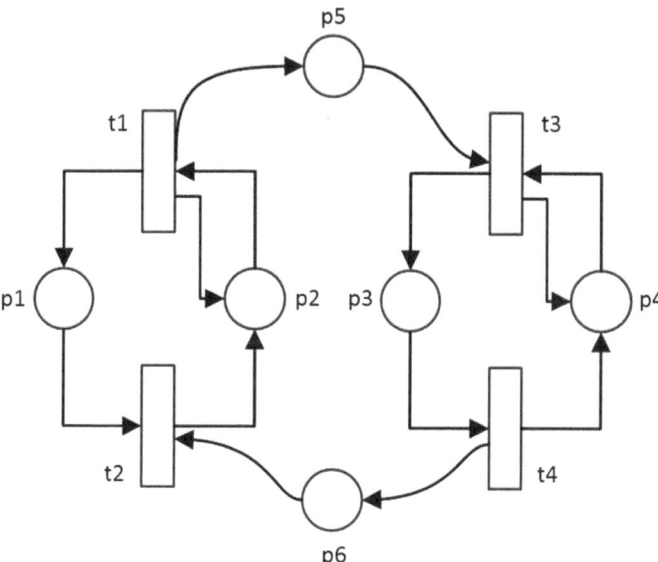

Fig. 8.3 A simple Petri net for automatically creating PDF

	p1	p2	p3	p4	p5	p6
t1	0	1	0	0	0	0
t2	1	0	0	0	0	1
t3	0	0	0	1	1	0
t4	0	0	1	0	0	0

Input adjacency matrix (Ai)

	p1	p2	p3	p4	p5	p6
t1	1	1	0	0	0	0
t2	0	1	0	0	0	0
t3	0	0	1	1	0	0
t4	0	0	0	1	0	1

Output adjacency matrix (Ao)

Fig. 8.4 Input and output adjacency matrices of the Petri net

With these two matrices, we can simply create a PDF by using the function
'**createPDF**'.

```
% Example-32: Testing automatic PDF creation
% Ai: input incidence matrix
Ai = [0 1 0 0 0 0; ...
      1 0 0 0 0 1; ...
      0 0 0 1 1 0; ...
      0 0 1 0 0 0];

% Ao: output incidence matrix
Ao = [1 0 0 0 1 0; ...
      0 1 0 0 0 0; ...
      0 0 1 0 0 0; ...
      0 0 0 1 0 1];

% optional: name for the PDF to be created
PDF_Filename = 'ex_32_pdf.m';

% optional: name for the Petri net
PN_name = 'Example 32 for automatic PDF creation';

% call the functio to create the PDF file
createPDF(Ai, Ao, PDF_Filename, PN_name);
```

createPDF will output the following PDF. Please note that all the transitions
and places are named in numerically increasing order.

```
% This PDF file is generated by "createPDF" function on
% On 09-Jun-2016    at 11:9:32
% PDF: ex_32_pdf.m

function [png] = ex_32_pdf()

png.PN_name = 'Example 32 for automatic PDF creation';
png.set_of_Ps = {'p1','p2','p3','p4','p5','p6'};
png.set_of_Ts = {'t1','t2','t3','t4'};
png.set_of_As = {'p2','t1',1,  't1','p1',1,  't1','p5',1,  ... % t1
                 'p1','t2',1,  'p6','t2',1,  't2','p2',1,  ... % t2
                 'p4','t3',1,  'p5','t3',1,  't3','p3',1,  ... % t3
                 'p3','t4',1,  't4','p4',1,  't4','p6',1,  ... % t4
                };
```

8.3.2 Example-33: A Large Cyclic Example

Figure 8.5 shows a cyclic Petri net. Let us assume that this cycle consists of a very large number of places and transitions; let us say that $n = 1000$. This means we have to start with a PDF that has 1000 place names and 1000 transition names, in addition to 2000 arcs. Coding such a large PDF by ourselves will be an impossible job to do! Luckily, with createPDF, creating a PDF will be a piece of cake!

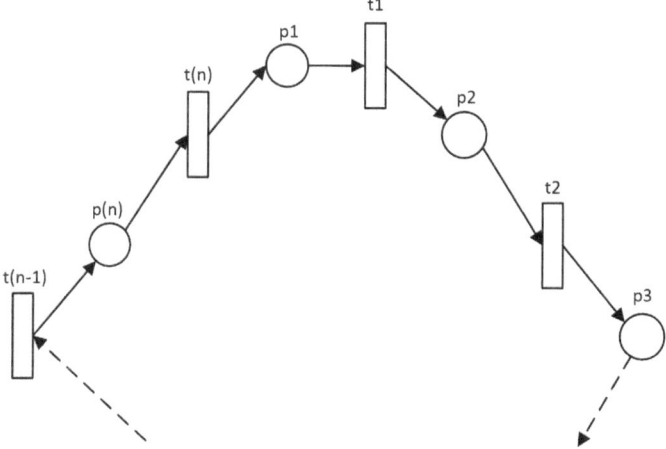

Fig. 8.5 A cyclic Petri net model for automatically creating PDF

```
% Example-33: Create a PDF that is impossible to create by hand
clear all; clc;

m = 1000; % number of places;
n = 1000; % number of transitions

% Ai: input incidence matrix
Ai = eye(m, n);

% Ao: output incidence matrix
Ao = zeros(m, n);
for i = 1: n-1,
    Ao(i, i+1) = 1;
end
Ao(n,1)=1;

% optional: name for the PDF to be created
PDF_Filename = 'ex_33_pdf.m';

% optional: name for the Petri net
PN_name = 'Example 33 (Cyclic PN) for automatic PDF creation';

% call the functio to create the PDF file
createPDF(Ai, Ao, PDF_Filename, PN_name);
```

A very large PDF file (named 'ex_33_pdf.m') will be created that has a 1211 lines, consisting 1000 place names, 1000 transition names, and 2000 arcs. Given below is the edited (shortened) version of the PDF:

```
% This PDF file is generated by "createPDF" function on
% On 09-Jun-2016    at 11:28:24
% PDF: ex_33_pdf.m

function [png] = ex_33_pdf()

png.PN_name = 'Example 33 (Cyclic PN) for automatic PDF creation';
png.set_of_Ps = {'p1','p2','p3','p4','p5','p6','p7','p8','p9',...
    'p10','p11','p12','p13','p14','p15','p16','p17','p18','p19',...
    ...
    ...

 'p990','p991','p992','p993','p994','p995','p996','p997','p998','p999
',...
            'p1000'};

png.set_of_Ts = {'t1','t2','t3','t4','t5','t6','t7','t8','t9',...
   't10','t11','t12','t13','t14','t15','t16','t17','t18','t19',...
    ...
    ...
 't992','t993','t994','t995','t996','t997','t998','t999',...
            't1000'};
png.set_of_As = {'p1','t1',1,  't1','p2',1,  ... % t1
                 'p2','t2',1,  't2','p3',1,  ... % t2
                 'p3','t3',1,  't3','p4',1,  ... % t3
                 'p4','t4',1,  't4','p5',1,  ... % t4
    ...
    ...

                 'p999','t999',1,  't999','p1000',1,  ... % t999
                 'p1000','t1000',1,  't1000','p1',1,  ... % t1000
              };
```

8.4 Interfacing with External Hardware

GPenSIM supports interacting with external devices through processor files (pre and post). In GPenSIM, places are assumed to be passive, and the transitions are assumed active. Thus, GPenSIM maps transitions to activate or deactivate external devices. Figure 8.6 shows how a transition can be used to control an external device. Specifically, the transition is to be used to switch on a light for 5 s:

- The pre-processor of the transition, as usual, first checks whether the transition can fire by going through (if any) additional conditions for firing. The pre-processor can also execute any commands coded in the pre-processor file. For example, the relevant command for the light switching example will be to **switch the light on**.
- Followed by the pre-processor, the firing transition enters sleeping in the 'firing_transition' queue; firing transition is just a passive time delay that will not consume CPU. **A firing transition sleeping during the firing time (5 s, in this example)** will automatically awaken when the firing is completed.
- When the firing is complete, the post-processor will be executed, executing all the commands on it; in the example, post-processor will switch off the light.
- Summary: (pre-processor switching on the light—transition sleeps (delayed) for 5 s—and the post-processor switching off the light): results in the light switched on for 5 s.

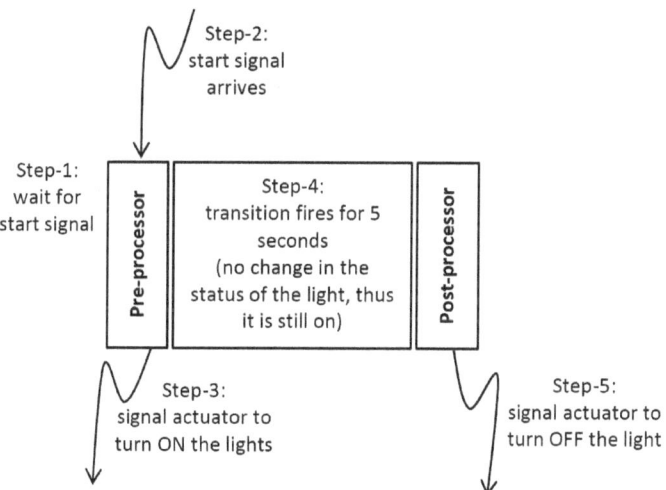

Fig. 8.6 Transition switching on an external actuator

IMPORTANT!
Remember to set the REAL_TIME flag to 'on' in the MSF:
    global_info.REAL_TIME = 1

8.4.1 Example-34: Switching Lights on a LEGO Mindstorm NXT Robot

This example is the same as the one described in example-06: firing alternatingly using binary semaphore. However, this time, we will use real hardware, a LEGO Mindstorm NXT robot, to alternatingly 'ON' and 'OFF' red and green lights. Of course, we need the following software and hardware installed in the system:

- The RWTH–Mindstorms NXT Toolbox
- LEGO Mindstorms NXT building kit 2.0; LEGO Mindstorms NXT firmware v1.26 or higher
- For Bluetooth wireless connection: Bluetooth 2.0 adapter recommended by LEGO
- For USB connections: Mindstorms NXT Driver 'Fantom' v1.02

The RWTH–Mindstorms NXT Toolbox (RWTH–Mindstorms NXT Toolbox 2017) is a software package available for the MATLAB environment that allows control of LEGO Mindstorm NXT robots from a personal computer; a LEGO robot can be controlled from a personal computer either through USB connection or through Bluetooth wireless connection.

By using GPenSIM on top of RWTH–Mindstorms NXT Toolbox, various discrete control applications (especially, supervisory control applications) can be developed. Figure 8.7 shows the various layers of software that are needed for developing applications for discrete robotic control.

Figure 8.8 shows the LEGO Mindstorms NXT robot setup with lights (red, green, and yellow) and switches ('buttons').

This is example, we are going to switch red and green color lights on the robot alternatingly for 2 s. The Petri net model is shown in Fig. 8.9, which is the same as the one in the example-06.

PDF (same as in example-06):

```
% Example-34:Real-Time Load balance (with LEGO NXT)
function [png] = loadbalance_pdf()

png.PN_name = 'Load Balancer';
png.set_of_Ps = {'p0', 'p1', 'p2'};
png.set_of_Ts = {'t1', 't2'};
png.set_of_As = {'p0','t1',1, 't1','p1',1,'p0','t2',1, 't2','p2',1};
```

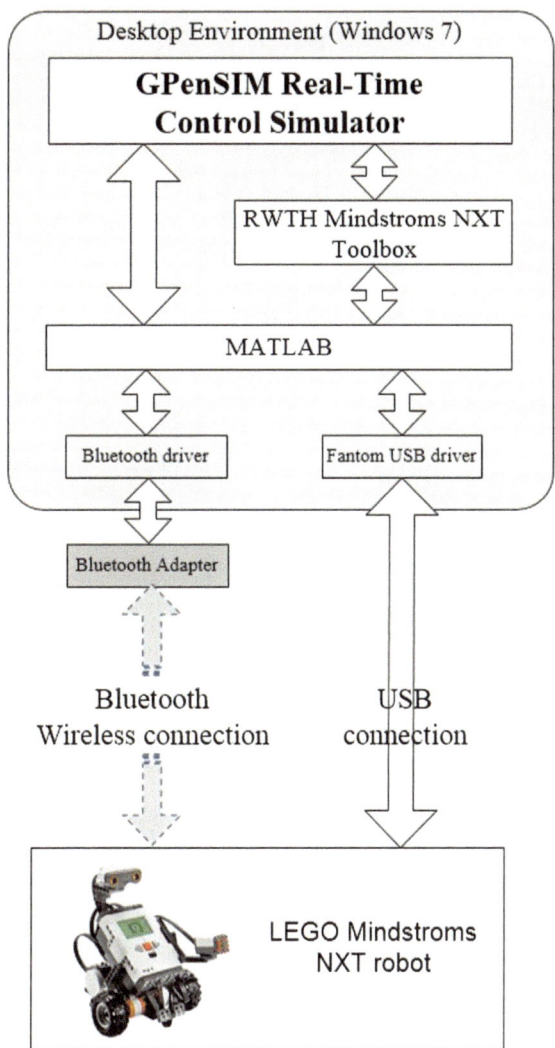

Fig. 8.7 Using LEGO NXT Mindstorms with GPenSIM

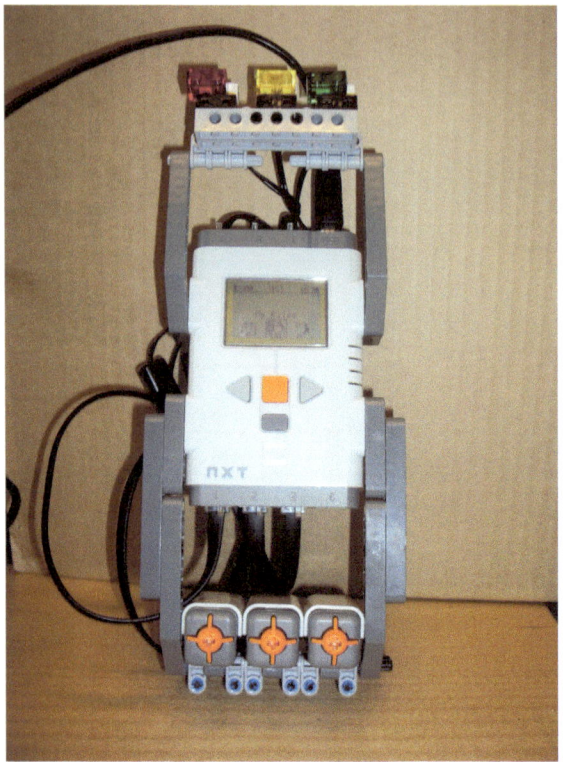

Fig. 8.8 LEGO NXT Mindstorms robot setup with lights and switches

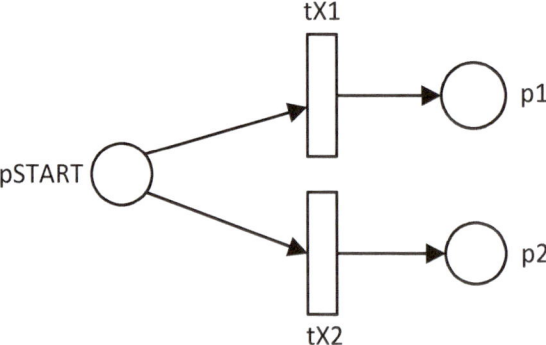

Fig. 8.9 Petri net for alternating firing

MSF: The only changes are the hardware initialization and real-time settings.

```
% Example-34: Rea-Time Load balance (with LEGO NXT)
clear all; clc;

global global_info
global_info.REAL_TIME = 1;      % set REAL_TIME flag
global_info.STOP_AT = current_clock(3)+[0 0 30]; % stop after 30 secs

global_info.semaphore = 't1'; % GLOBAL DATA: binary semafor

init_TL_NXT();  % Initialize LEGO H/W

png = pnstruct('loadbalance_pdf');

dyn.m0 = {'p0',9};
dyn.ft = {'t1',2, 't2',2};
pni = initialdynamics(png, dyn);

sim = gpensim(pni);
plotp(sim, {'p1', 'p2'});

close_TL_NXT(); % Important: clear memory (and hardware) after use
```

init_TL_NXT (initialize LEGO NXT hardware):

```
% Example-34: Rea-Time Load balance (with LEGO NXT)
function [] = init_TL_NXT()
% initialize the lights in NXT

global global_info

% initializ NXT
warning('off', 'MATLAB:RWTHMindstormsNXT:noEmbeddedMotorControl');

COM_CloseNXT all
hNXT = COM_OpenNXT('bluetooth.ini');        % look for USB devices
COM_SetDefaultNXT(hNXT);     % sets global default handle

global_info.NXT_handle = hNXT;

global_info.lightRED = MOTOR_A;
global_info.lightGREEN = MOTOR_C;

NXT_PlayTone(440, 500); % just play a music to indicate ready
```

Close_TL_NXT (for clearing the memory and hardware at the completion of simulations):

```
function [] = close_TL_NXT()
% function [] = close_TL_NXT()

global global_info

SwitchLamp(global_info.lightRED, 'off');
SwitchLamp(global_info.lightGREEN, 'off');

% Never forget to clean up after your work!!!
COM_CloseNXT(global_info.NXT_handle);
```

COMMON_PRE:

```
function [fire, trans] = COMMON_PRE(trans)
% COMMON_PRE file codes the enabling conditions
% Here, the current value of the semaphore
% indicate which transition can fire.
%
% After firing, the fired transition must
% set the value of semaphore to the other
% transition: this is done in the COMMON_POST
% file.
global global_info

if strcmpi(trans.name, 't1'),
    fire = strcmp(global_info.semaphore, 't1');
    if (fire), SwitchLamp(global_info.lightRED, 'on'); end

else strcmp(trans.name, 't2'),
    fire = strcmp(global_info.semaphore, 't2');
    if (fire), SwitchLamp(global_info.lightGREEN, 'on'); end
end
```

COMMON_POST:

```
function [] = COMMON_POST(transition)
% COMMON_POST file codes the post firing actions.
% Here, right after firing, the fired transition
% set the value of semaphore to the other
% transition so that the other one fires next
global global_info

if strcmp(transition.name, 't1'),
    SwitchLamp(global_info.lightRED, 'off');
    global_info.semaphore = 't2'; % t1 releases semafor to t2
else % transition.name 't2'
    SwitchLamp(global_info.lightGREEN, 'off');
    global_info.semaphore = 't1'; % t2 releases semafor to t1
end
```

8.5   Interfacing with Graph Algorithms Toolbox

Being a directed bipartite graph, Petri nets can be analyzed for properties based on the graph algorithms, e.g., the shortest distance between any two nodes and the existence of cycles. To run graph algorithms, we have to change the bipartite Petri net into a directed homogeneous graph (digraph).

8.5.1   Example-35: Converting a Bipartite Petri net into a Homogeneous Digraph

In this example, we are going to convert a bipartite Petri net into a homogeneous digraph so that some graph algorithms can be run on it. Consider the Petri net shown in Fig. 8.10.

Let us convert the Petri net into a homogeneous digraph. The resulting digraph is shown in Fig. 8.11.

The GPenSIM function 'convert_PN_V' is the one that performs the conversion. The code for the conversion is given below:

MSF:

```
% Example-35: Converting a PN into a digraph
pns = pnstruct('pn2dg_pdf');
disp('The incidence matrix of PN: ');
disp(pns.incidence_matrix);

V = convert_PN_V(pns); % Transform the Petri net into a homogeneous digraph

% the homogeneous digraph V can be
% fed into graph algorithms, e.g. dijsktra
disp(' ');
disp('The adajacency matrix of the digraph: ');
disp(V.A);
```

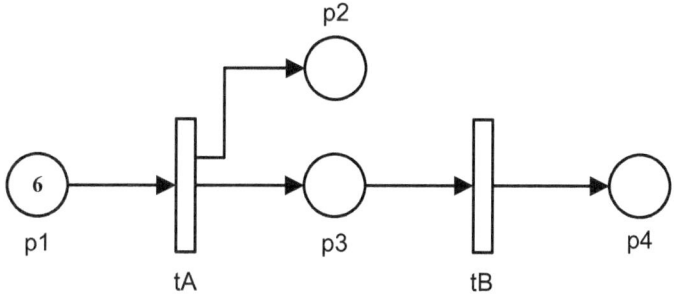

Fig. 8.10 A simple Petri net for conversion into a homogeneous digraph

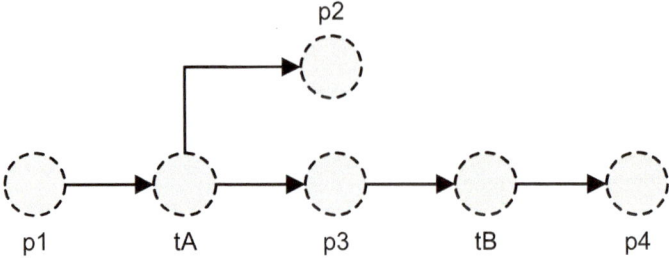

Fig. 8.11 The digraph obtained from the Petri net shown in Fig. 8.10

The output of the code is:

```
The incidence matrix of PN:
    1     0     0     0     0     1     1     0
    0     0     1     0     0     0     0     1


The adajacency matrix of the digraph:
    0     0     0     1     1     0
    0     0     0     0     0     1
    1     0     0     0     0     0
    0     0     0     0     0     0
    0     1     0     0     0     0
    0     0     0     0     0     0
```

The incidence matrix of the Petri net has two parts as shown in Fig. 8.12.
However, the adjacency matrix of the digraph has four parts, as shown in Fig. 8.13.

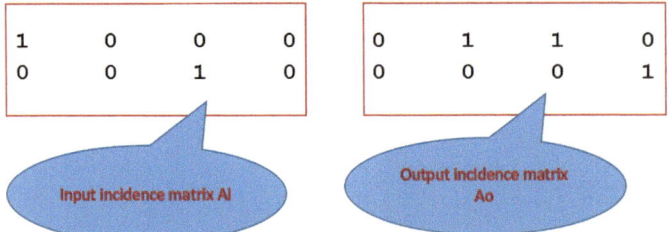

Fig. 8.12 Incidence matrix of a Petri net in GPenSIM

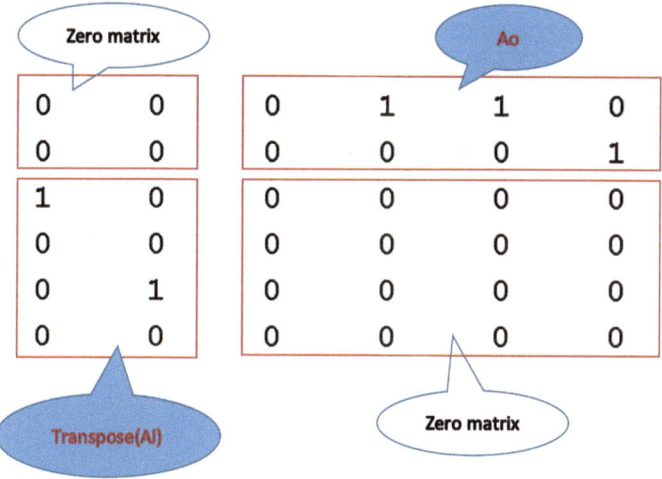

Fig. 8.13 Adjacency matrix of the resulting homogeneous digraph

References

PIPE2. (2017). Available: http://pipe2.sourceforge.net/.
PNML reference cite. (2017). Available: http://www.pnml.org/.
RWTH–Mindstorms NXT Toolbox. (2017). Available: http://www.mindstorms.rwth-aachen.de/.

Chapter 9
Structural Invariants

This chapter presents the functions that are available in GPenSIM for finding the structural invariants (e.g., siphons, traps, place invariants (P-invariants), and transition invariants (T-invariants). Structural invariants (*aka* net invariants) are the structural properties of a Petri net that depend only on the static (topological) structure; structural invariants are independent of the Petri net's markings.

9.1 GPenSIM Functions for Structural Properties

The Petri net Control Toolbox (PNCT) developed at the University of Cagliari offers a simple and crude functionality for analyzing structural properties. GPenSIM extends the functionality to provide a more comprehensible information. It is easy to integrate PNCT with GPenSIM: All we have to do is to transform the static Petri net structure (PN) into input incidence matrix (A^- or D^-) and output incidence matrix (A^+ or D^+).

```
function [Pre_A, Post_A, D] = gpensim_2_PNCT(A)
% function [Pre_A, Post_A, D] = gpensim_2_PNCT(A)
%
% convert GPenSIM Incidence Matrix (A) structure into
%     Cagliari "Petri net Control Toolbox" incidence matrices
%
%   Reggie.Davidrajuh@uis.no (c) Version 10.0 (c) July 2017
%%%%%%%%%%%%%%%%%%%%%%%%%%%%%%%%%%%%%%%%%%%%%%%%

Ps = size(A, 2)/2;  % Number of places
Pre_A  = A(:, 1:Ps)';       % Pre  places
Post_A = A(:, Ps+1:2*Ps)';  % Post places
D = Post_A - Pre_A;         % Incidence Matrix
```

© The Author(s) 2018
R. Davidrajuh, *Modeling Discrete-Event Systems with GPenSIM*, SpringerBriefs in Applied Sciences and Technology, https://doi.org/10.1007/978-3-319-73102-5_9

Table 9.1 Structural invariants and their properties

Structural invariant	Properties
Place invariants (P-invariants)	Place invariants are the set of places whose weighted token sum remains constant for all possible markings
Transition invariants (T-invariants)	Transition invariants are a set of firings that will cause a cycle in the state space, meaning we will come back to the original state (markings)
Traps	Traps are a set of places which if become marked will always remain marked for all reachable markings of the net
Siphons	Siphons are a set of places, which if they become empty of tokens, will always remain empty for all reachable markings of the net

Table 9.1 presents some of structural invariants:

9.1.1 Example-36: Finding Siphons and Minimal Siphons

This example uses the same Petri net as in the example-09. The Petri net model is shown in Fig. 9.1. We will see soon that the Petri net has a single siphon ('p1') and a single trap ('p2').

The functions **'siphons_minimal'** and **'siphons'** return a matrix of place indices, in addition to printing the siphons on the screen.

```
% Example-36: Siphon example
pns = pnstruct('siphons_pdf');
SM = siphons_minimal(pns);
disp('Minimal siphons (in matrix form): ');
disp(SM);

S = siphons(pns);
disp('Siphons (in matrix form): ');
disp(S);
```

Fig. 9.1 Finding siphons

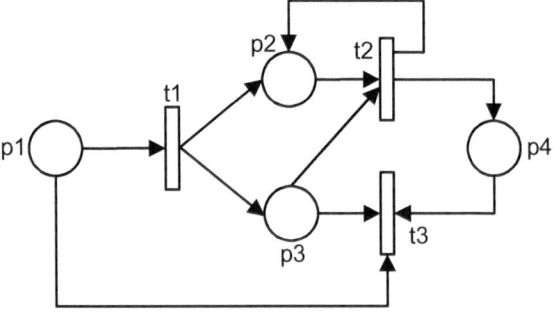

The minimal siphons and siphons are:

```
Minimal siphons in this net:
{p1}
Minimal siphons (in matrix form):
     1  0  0  0

Siphons in this net:
{p1}
{p1,p2}
{p1,p3}
{p1,p2,p4}
{p1,p3,p4}
Siphons (in matrix form):
    1      0      0      0
    1      1      0      0
    1      0      1      0
    1      1      0      1
    1      0      1      1
```

Analysis of the results: The results show that '**p1**' is a siphon. This is correct as **t1** and **t3** can fire a number of times to remove (siphon) all the tokens in **p1**. Once **p1** becomes empty, it will remain as empty as there is not any transition that will deposit tokens into **p1** as it is a source.

9.1.2   Example-37: Finding Traps and Minimal Traps

The functions for finding traps and minimal traps are very similar to finding siphons. Using the same net from the previous example (example-36), we will find traps and minimal traps. Functions **traps_minimal** and **traps** return a matrix of place indices and also print the traps on the screen.

```
% Example-37: Traps example
pns = pnstruct('traps_pdf');
TM = traps_minimal(pns);
disp('Minimal traps (in matrix form): ');
disp(TM);

T = traps(pns);
disp('Traps (in matrix form): ');
disp(T);
```

The traps and minimal traps are:

```
Minimal traps in this net:
{p2}
Minimal traps (in matrix form):
     0    1    0    0

Traps in this net:
{p2}
Traps (in matrix form):
     0    1    0    0
```

Analysis of the results: The results also show that '**p2**' is a trap, which is correct. This is because, if **p2** gets any token by firing of **t1**, **p2** will never lose the tokens as the one that removes the tokens (**t2**) will always put it back into **p2**.

9.1.3   Example-38: Finding P-Invariants and T-Invariants

GPenSIM functions **pinvariant** and **tinvariant** can be used to find the invariants. In this example, we will find the P- and T-invariants of the Petri net shown in Fig. 9.2.

PDF:

```
% Example-38: Finding P-invariants and T-invariants
function [png] = ptinvar_pdf()
png.PN_name = 'P-invariants and T-invariants';
png.set_of_Ps = {'p1', 'p2', 'p3', 'p4'};
png.set_of_Ts = {'t1', 't2', 't3'};
png.set_of_As = {...
    'p4','t1',1, 't1','p1',1, 't1','p2',1, ...   %t1
    'p1','t2',1, 't2','p3',1, ...                %t2
    'p2','t3',1, 'p3','t3',1, 't3','p4',1, ...   %t3
    };
```

MSF:

```
% Example-38: P/T Invariants
pns = pnstruct('ptinvar_pdf');
PI = pinvariant(pns);
TI = tinvariant(pns);
```

Fig. 9.2 Petri net for testing P- and T-invariants

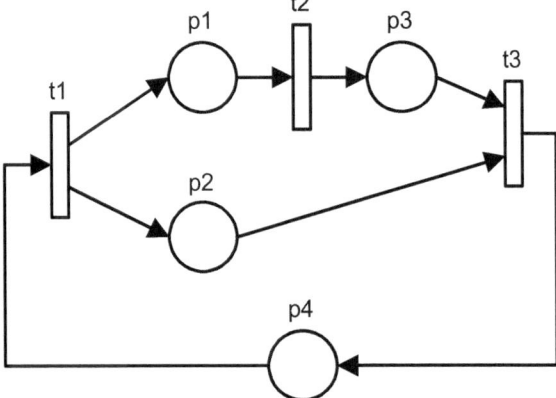

Simulation results:

```
P-invariants:
{p2,p4}
{p1,p3,p4}

T-invariants:
{t1,t2,t3}
```

Analysis of the results: The results also show:

P-invariants: (1) The weighted sum of the tokens in **p2** and **p4** is a constant, and (2) the weighted sum of the tokens in **p1**, **p3**, and **p4** is also a constant.

T-invariants: No matter what the state is, if the transitions **t1**, **t2**, and **t3** are fired, then we return to the original state. This property can be verified by 'firing sequence' (Sect. 4.2).

Bibliographical Remarks

For an extensive coverage of the structural invariants and their applications on supervisory control, the following texts are suggested: Iordache and Antsaklis (2007) and Moody and Antsaklis (1998).

References

Iordache, M., & Antsaklis, P. J. (2007). *Supervisory control of concurrent systems: A Petri net structural approach*. Berlin: Springer Science & Business Media.

Moody, J. O., & Antsaklis, P. J. (1998). *Supervisory control of discrete event systems using Petri nets*. Dordrecht: Kluwer Academic Publishers.

Appendix
Frequently Used GPenSIM Functions

Some of the frequently used GPenSIM functions are given below, in the alphabetical order:

Function	Purpose
check_valid_place	Verify whether the given place is a valid one
check_valid_transition	Verify whether the given transition is a valid one
convert_PN_V	Convert bipartite Petri net into a homogenous directed graph with an adjacency matrix representing the connections between the homogenous nodes
cotree	Generate coverability tree of a Petri net
cotreei	Generate coverability tree of a Petri net with inhibitor arcs
createPDF	Create a PDF file from the input adjacency matrices
current_clock	Returns the real-time clock in [hour min sec] format
current_time	Returns the current value of the simulated time
cycles	Detect all the elementary circuits (cycles) in a Petri net
extractt	Get the start time and stop time of every transition firing
firingseq	To fire a pre-assigned sequence of transitions
get_firingtime	Returns the firing time of a transition
get_inputplace	Returns the input places of a transition
get_inputtrans	Returns the input transitions of a place
get_outputplace	Returns the output places of a transition
get_outputtrans	Returns the output transitions of a place
get_place	Returns the details of a place (name and number of tokens)
get_priority	Return the current priority of a transition
get_tokCT	Returns the creation time of a token (identified by tokID)
get_token	Get complete information about the token in a place
get_tokens	Get complete information about the set of tokens in a place
gpensim	Run simulations on a Petri net with initial dynamics
gpensim_ver	Prints the current version of GPenSIM

(continued)

© The Author(s) 2018
R. Davidrajuh, *Modeling Discrete-Event Systems with GPenSIM*, SpringerBriefs in Applied Sciences and Technology, https://doi.org/10.1007/978-3-319-73102-5

(continued)

Function	Purpose
`initialdynamics`	Add the initial dynamics to the static Petri net structure
`is_enabled`	Verify whether a transition is enabled
`is_eventgraph`	Verify whether a Petri net is an event (marked) graph
`is_firing`	Verify whether a transition is firing
`is_stronglyconn`	Verify whether a Petri net is strong connected
`matrixD`	Compute the incidence matrices D^-, D^+, and D
`mincyctime`	Find the minim cycle time of an event graph
`nplaces`	Total number of places in a Petri net
`ntokens`	Number of tokens residing in a place
`ntrans`	Total number of transitions in a Petri net
`occupancy`	Find the overall firing times (in percentage) of the transitions
`pinvariant`	P-invariants of a Petri net
`plotp`	From the simulation results, plot the figure showing how the tokens in the places vary with time
`pname`	Name of the place identified by the place_index
`pnclass`	Find the Petri net class
`pnml2gpensim`	Generate GPenSIM files (MSF, PDF, and templates for COMMON_PRE and _POST) from PNML file
`pnstruct`	Create the static Petri net structure from the PDF file.
`priorcomp`	Compare the current priorities of two transitions
`priordec`	Decrease the priority of a transition by one
`priorinc`	Increase the priority of a transition by one
`priorset`	Set the priority of a transition to an integer value
`prnss`	From the simulation results, print the state space diagram
`prnstate`	Print the current state
`rt_clock_string`	Returns the real-time clock [hour min sec] as a text string for display
`siphons_minimal`	Siphons of a Petri net
`stronglyconn`	Finding strongly connected components (SCC) in a Petri net
`timesfired`	Number of times a transition has already fired
`tinvariant`	T-invariants of a Petri net
`tname`	Name of the transition identified by the transition_index
`traps_minimal`	Traps of a Petri net

Index

© The Author(s) 2018
R. Davidrajuh, *Modeling Discrete-Event Systems with GPenSIM*, SpringerBriefs in
Applied Sciences and Technology, https://doi.org/10.1007/978-3-319-73102-5